NITROGEN MANAGEMENT IN IRRIGATED AGRICULTURE

Nitrogen Management in Irrigated Agriculture

Roy S. Rauschkolb
Department of Soil and Water Science
University of Arizona

Arthur G. Hornsby
Department of Soil and Water Science
University of Florida

New York   Oxford
OXFORD UNIVERSITY PRESS
1994

Oxford University Press

Oxford   New York   Toronto
Delhi   Bombay   Calcutta   Madras   Karachi
Kuala Lumpur   Singapore   Hong Kong   Tokyo
Nairobi   Dar es Salaam   Cape Town
Melbourne   Auckland   Madrid

and associated companies in
Berlin   Ibadan

Published by Oxford University Press, Inc.
200 Madison Avenue, New York, New York 10016

Library of Congress Cataloging-in-Publication Data
Rauschkolb, Roy S.
Nitrogen management in irrigated agriculture / Roy S. Rauschkolb
and Arthur G. Hornsby.
p.   cm.   Includes bibliographical references (p.   ) and index.
ISBN 0-19-507835-7
1. Irrigation farming. 2. Nitrogen in agriculture—Management.
3. Crops and nitrogen. I. Hornsby, Arthur G. II. Title.
S619.N57R38 1994
631.8'4—dc20   93-37138

9 8 7 6 5 4 3 2 1

Printed in the United States of America
on acid-free paper

To Joan and Kathie
for sharing our lives
and our dreams

Preface

Nitrogen constitutes about 79 percent by volume of the earth's atmosphere, and the amount present in the soil, water, and atmosphere has been estimated at approximately 10^{15} metric tons. However, in an agricultural plant production system, it is still common for available forms of soil nitrogen to be deficient for maximum plant growth. Consequently, nitrogen fertilizers are essential for the economic production of food and fiber. The potential for nitrogen pollution of receiving waters arises because the movement of nitrates below the root zone is inevitable. In order to minimize unwanted accumulations or entry of nitrogen into various sectors of the environment we must learn how to better manage our manufactured and naturally occurring nitrogen resources.

The best management practice for controlling man-induced nitrate accumulation in receiving waters is the management of nitrogen and water inputs on the land surface. For agricultural systems this will require on-farm management. Implementing on-farm control measures in irrigated agriculture should take into account the great diversity that exists in managerial ability, cultural practices, and site conditions. Implementation of control measures will require field personnel who understand soil, water, and plant relationships to guide the farmer in the use of techniques that provide flexibility in their application and ease of implementation and are adaptable to a wide range of site conditions and managerial abilities.

Agriculture does not exist independently from all other segments of society. Consequently, the agriculturist shares the concerns of others with regard to conservation of energy, land, and water resources and the impact of human activities on the environment. Furthermore, it is in agriculture's own best interest for sustained food and fiber production to maintain an environment compatible with the needs of others.

R.S.R.
A.G.H.

Contents

nitrogen forms and processes within the system. The relationships between transformations, loss pathways, and forms of nitrogen are described in the context of nutrient use efficiency.

Climate is an important factor that must be considered when managing cropping systems. The environmental variables, precipitation and temperature, are described in relation to their impact on nitrogen management.

PART II.   Management Variables

Fertilizer placement relative to crop roots impacts crop yields and fertilizer losses from the root zone. Cropping practices, soil properties and irrigation systems may determine how fertilizers are applied. Integration of these factors is described.

Availability of application equipment may determine how fertilizer nitrogen is applied and which forms of nitrogen are used. Factors affecting equipment selection and application uniformity are described.

Matching fertilizer application rate to plant demand is essential for both crop production and protection of groundwater. Concepts of efficient nutrient management in light of system variables are discussed.

The chemical and physical characteristics of different nitrogen fertilizers influence plant uptake and loss pathways. Volatilization, leaching and plant uptake are described as a function of nitrogen source and system variables.

Management of irrigation water influences the movement and use efficiency of nitrogen in the soil. Irrigation method and frequency are described in relation to system variables.

Timing of fertilizer applications is an essential management technique for improving efficiency of nitrogen use. The relationships between system variables and timing of applications are described.

NITROGEN MANAGEMENT IN IRRIGATED AGRICULTURE

1
Introduction

Events of the recent past have been moving agriculture rapidly toward the point where it can no longer be concerned with production of food and fiber alone. One of the principal reasons for the change in attitudes has been the Federal Water Pollution Control Act Amendment of 1972, commonly referred to as Public Law 92-500. Within this law, Section 208 deals with area-wide water quality management planning. It requires each state to develop waste water management strategies. Specific outputs resulting from Section 208 planning include a regulatory program to control or treat all point and nonpoint pollution sources, including accumulated levels of pollution. This represents only one of many outputs that are expected from Section 208 planning. It is the one that is especially important with respect to nitrogen management in irrigated agriculture.

Another factor having an impact on nitrogen management is the cost of energy. As fossil fuels become more costly and less available, the cost of producing fertilizer nitrogen will substantially increase. This will lead the grower to more efficient utilization of fertilizer resources. At present this does not appear to be a particularly strong motivating force for achieving efficiency. Furthermore, demand for food is going to increase as population increases, and since there is no alternative to increasing food production to meet the demand it seems likely that the increased costs of production when they do occur will have to be passed on to the consumer.

According to Stout (unpublished data, 1971), the on-farm requirement for nitrogen to maintain the average diet of the U.S. population is about 82 kilograms of nitrogen per year per capita. This also includes inefficiencies and various losses inherent to the agricultural food production system. Based on 1990 population, this means about 20.5 million metric tons of farmstead nitrogen will be needed to maintain the average diet. Approximately 40 percent of this nitrogen demand is being met by manufactured fertilizer nitrogen. The remainder of the nitrogen demand is made up by biological nitrogen fixation, recycling released from organic forms, and natural chemical fixation.

Improvement in efficiency of nitrogen fertilizer use and reduction of losses are certain to occur. These will be compensated by increased population, so the farm-site nitrogen demand may remain fairly steady for many years. The percentage of farm-site nitrogen requirement met by commercial fertilizers will, in all likelihood, increase. Although the quantity of fertilizer nitrogen seems quite large, it is still small in comparison to the total amount of nitrogen present in the earth's surface and atmosphere. Based on various estimates of the amount of nitrogen being fixed on a global basis, it appears that less than 1 millionth of a percent of the total nitrogen is being recycled annually.

Regardless of the vast quantity of nitrogen that is present in the soil, water, air, plant, and animal system, the general case in an unfertilized field is for plants to be nitrogen deficient. Consequently, there is need for additional nitrogen in order to produce the food and fiber needed by our society.

The use of nitrogen fertilizers also has beneficial effects with respect to utilization of our land and water resources. In an article on the environmental benefits of intensive crop production, Barrons (1971) documented a decrease in the area used for plant production from approximately 148 million hectares (365 million acres) to about 117 million hectares (290 million acres) in the period from 1930 to 1970. During the same period, the population in the United States grew from approximately 120 million to slightly more than 200 million. The increased food demand was met by utilizing less land area for production of food and fiber. In the same article, Barrons compared the area required for the production of 17 crops in the period 1938 to 1940 with that required in 1968–1970 for these same crops. Although other factors, such as improved management, improved varieties, and pest control, played an important role in the increased production, the widespread adoption of nitrogen fertilizers had a major impact on the increased production per unit land area or per unit input of water.

Consequently, these factors have substituted for increased development of water supplies and arable land. The area saved, as indicated by Barrons, was approximately 118 million hectares (292 million acres). According to the 1987 agricultural census there were approximately 114 million hectares of harvested cropland (U.S. Department of Commerce, 1990). These data continue to support the conclusions that Barrons made. Increased inputs and production per unit area are substituting for added land in production and are adequate to meet the increased demand for food by the 1990 population of 235 million people in the United States. Some food is imported from other countries, which would increase the actual area used for food production, but in fact slightly increase the area involved since there are approximately compensating exports and imports of agricultural products to the United States.

In spite of the many benefits that can be ascribed to the use of nitrogen fertilizers, there is ample evidence to indicate that inefficient utilization of such fertilizers can lead to pollution of surface and groundwater supplies. Nitrate-N concentrations reported in Table 1-1 attest the potential for groundwater contamination. Only two of fifty states were determined to have no reported wells with concentrations of nitrate-N over 10 mg/l. Comparisons of the data between states indicate that in some states the data are skewed toward increased nitrate levels in the groundwater. Over 10 percent of the states have nitrate-N concentrations reported to be in excess of the 10 mg/l drinking water standard. While the source of nitrate-N in the wells sampled is not known, those states with high reported concentrations represent intense agriculture and/or urban development.

Table 1-1.   Summary of nitrate-nitrogen concentrations in ground water, by
              state[a].

State	Number of wells sampled	Percentage of wells for which maximum nitrate-N concentrations fell within indicated range (mg/l)			
		0-0.2	0.21-3.0	3.1-10	>10
Alabama	244	47.1	45.5	7.4	0.0
Alaska	1,305	60.9	33.9	2.8	2.4
Arizona	4,164	12.1	49.7	24.4	13.9
Arkansas	2,436	49.1	38.5	8.5	3.9
California	2,732	21.9	45.4	22.5	10.1
Colorado	5,492	33.8	43.3	17.2	5.7
Connecticut	348	33.6	49.7	14.4	2.3
Delaware	165	34.5	30.9	25.5	9.1
Florida	3,140	71.5	24.2	2.3	2.0
Georgia	1,137	66.7	28.5	4.3	0.5
Hawaii	164	15.9	75.0	9.1	0.0
Idaho	1,806	33.3	52.0	12.9	1.7
Illinois	359	56.0	30.1	5.6	8.4
Indiana	650	55.4	33.4	9.7	1.4
Iowa	4,088	44.9	36.7	13.4	5.0
Kansas	1,140	17.0	28.8	34.2	20.0
Kentucky	3,227	36.5	46.2	13.0	4.2
Louisiana	3,177	78.3	19.4	1.8	0.6
Maine	147	50.3	35.4	12.2	2.0
Maryland	1,521	40.9	30.4	22.0	6.8
Massachusetts	414	42.3	52.2	4.3	1.2
Michigan	1,108	79.1	17.1	2.8	1.1
Minnesota	1,655	39.1	40.7	10.9	9.3
Mississippi	1,701	76.5	21.7	1.6	0.2
Missouri	2,165	64.2	27.2	6.6	2.1
Montana	2,812	43.4	45.1	7.7	3.8
Nebraska	2,326	18.0	49.3	23.4	9.3
Nevada	465	46.2	45.4	7.5	0.9
New Hampshire	69	66.7	29.0	2.9	1.4
New Jersey	1,385	63.0	25.6	10.0	1.4
New Mexico	4,685	38.4	48.9	9.8	2.9
New York	2,491	28.9	38.0	29.3	11.0
North Carolina	908	72.1	22.0	5.1	0.8
North Dakota	7,387	22.4	68.5	4.4	4.6
Ohio	339	61.7	29.8	5.9	2.6
Oklahoma	1,724	23.0	41.2	24.1	11.8
Oregon	685	57.1	36.4	5.4	1.2
Pennsylvania	4,326	31.1	38.7	24.4	5.9
Puerto Rico	79	16.5	48.1	32.9	2.5
Rhode Island	171	17.0	38.0	8.8	36.3
South Carolina	557	69.3	26.6	3.4	0.7
South Dakota	1,996	49.2	35.9	8.2	6.7
Tennessee	109	65.1	29.4	4.6	0.9
Texas	36,196	76.5[b]	76.5[b]	14.1	9.4
Utah	3,301	39.1	50.4	8.4	2.0
Vermont	73	52.1	41.1	5.5	1.4
Virginia	762	70.7	25.9	2.6	0.8
Washington	1,158	38.3	41.1	18.6	4.3
West Virginia	954	68.6	25.9	5.0	0.5
Wisconsin	2,727	40.1	41.3	15.1	3.6
Wyoming	1,477	47.9	40.7	7.6	3.8
Total or percentage	123,656	41.6[c]	41.1[c]	13.2	6.4

Source: Madison and Brunett, 1984.
[a] Percentages for each state may not add to 100 percent because of independent
rounding; Source: Data from samples collected and analyzed by the U.S. Geological
Survey and Texas Department of Natural Resources over a 25 year period.
[b] Analysis of samples in Texas not reported for 0-0.2 ml/l range, but rather 0-3.0
ml/l range.
[c] Excluding data from Texas.

Similarly, some pollution also is attributable to the natural presence of nitrogen in vast quantities in our surroundings.

Considerable debate has been carried out in scientific circles regarding the best approach to minimizing the unwanted entry of nitrogen into the environment from various sources. The debate has centered around two possible mechanisms for minimizing nitrogen pollution potential: (1) treatment of discharge waters, or (2) control of inputs in order to minimize nitrogen output. For domestic and industrial water discharge, where the discharge occurs at an identifiable point and where the volumes are relatively small, it appears that treatment of waste water is the most appropriate method to minimize nitrogen pollution potential. In an agricultural system the situation is much different, since there are two distinct types of discharges and the discharge volumes for both are very much greater. One type of discharge is an identifiable point source, such as might occur in discharge of effluent from a drainage tile. The other type is a diffuse nonpoint discharge, which is water percolating below the root zone to an underground water table or returning to surface waters by subsurface flow. Although the technology exists for treatment of agricultural point source discharges, it is generally agreed that the volumes of water that must be handled make this an economically impractical alternative. Because of the diffuse nature of percolating water and the virtual impossibility of collecting such water for treatment before it reaches the deep water table, it is generally agreed that treatment of nonpoint sources of nitrogen pollution is not a viable alternative. Consequently, education of users on best management practices to minimize the amount of residual soil nitrogen available for leaching is the method most likely to accomplish a reduction in nitrate pollution potential.

Monitoring of nitrogen concentrations in percolating waters has been suggested as one technique for evaluating the nitrogen pollution potential of such waters. One of the problems inherent in using this technique is the extreme variability that occurs with respect to nitrogen concentrations in soils. In a California study, funded by the Research Applied to National Needs (RANN) Division of the National Science Foundation (Pratt 1979a), it was found that it would require anything from 5 to 20 holes per site (with 9 samples per hole) to reach a level of precision within 20 percent of the mean. The number of holes required varied with irrigation method, crop type, and soil texture. Even if one could afford the cost required to reach the indicated level of precision, that level of precision might not be satisfactory in determining whether excessive nitrogen inputs had been made or not. That level of precision also is believed by many to be no better than what might be achieved through field management of nitrogen and water. Furthermore, it is even expected that some field control measures will actually result in a *reduction*, rather than an increase, in production costs.

For these reasons it appears that the least cost and most effective method for minimizing nitrogen pollution from irrigated agriculture is to employ management techniques that result in the greatest crop output for each unit of nitrogen and water input. Variations that can occur in the management strategies may be employed to operate at peak efficiency attainable for a given set of conditions. Using this approach, peak operational efficiency (Best Management Practices) becomes the mechanism that will lead to greater nitrogen use efficiency, with a concomitant reduction in nitrogen pollution potential.

The purpose of this book is to define the principal factors that delimit the agricultural plant production system and to provide the basis for knowing where, when, and how different nitrogen and water management techniques can be employed to control nitrogen use efficiency and nitrogen-related water pollution. In Part I, the production system variables discussed are soil, crop, irrigation method, nitrogen cycling, and environment (SCINE). These variables are presented as if they operated independently in the agricultural system. Although it is recognized that they are not entirely independent of one another, their influence is sufficiently independent in the system that they warrant separate discussion. In combination, these variables determine the conditions that exist for a specific site, which, in turn, influence the management techniques used to minimize or eliminate unwanted nitrogen entry into receiving waters.

Part II deals with those management variables that can be employed within a given set of system variables to increase nitrogen use efficiency and reduce pollution potential. The management variables are placement, equipment, rate, source, irrigation management, timing and energy (organic matter). Their acronym, PER SITE, assists in recognition of the management variables that can be applied to a given set of conditions. It also emphasizes the importance of considering those site-specific conditions that determine how each of the management techniques will be employed. Another important feature of these management techniques is that they are greatly interdependent. Their interrelationships will also be discussed.

This book is presented as a compendium of those factors that are viewed as having major importance in agricultural systems. Within those systems, practices are stressed that can be employed to improve nitrogen management in irrigated agriculture in order to sustain agricultural production of food and fiber and to reduce nitrogen pollution potential in areas where it is of concern.

PART I
SYSTEM VARIABLES

2
Soils

Soils are very complex systems that have developed over time as a result of the principal soil-forming factors as elucidated by Hans Jenny many years ago. Those factors are climate, topography, parent material and biological activity along with time. The physical and chemical variability of soils arises from the great variations in soil forming factors found from one location to another. These, in turn, influence plant development and biological activity of soil. The purpose of this discussion is to provide an awareness of the characteristics of soils that are influencing plant growth and changes in the nitrogen pools in the rhizosphere. This perspective of various soil characteristics allows more informed decisions with respect to nitrogen management.

PHYSICAL CHARACTERISTICS

The four major physical characteristics influencing plant growth and nitrogen transformation in soils are structure, texture, uniformity and depth. These factors combine to make up a given soil management unit and govern in part the yield potential of a crop. A management unit may be defined as the area for which all cultural practices are performed in the same way. How these physical characteristics influence plant growth is illustrated in Fig. 2-1.

Structure

The structure of the soil plays a major role in governing root growth and the movement of water and air. The solid particles of the soils are made up primarily of inorganic minerals whose particle sizes impart a characteristic to soils known as *texture*. Organic solid particles contain varying amounts of relatively stable end products of decomposition, organic materials in various stages of decomposition, and living organisms such as insects, microorganisms, and plants. Through a variety of chemical and physical forces, these particles may take on different arrangements referred to as *soil structure*. Granular structure (Fig. 2-2) occurs when the individual particles are combined to form larger particles through the binding forces between clay minerals, sand, and organic matter. The

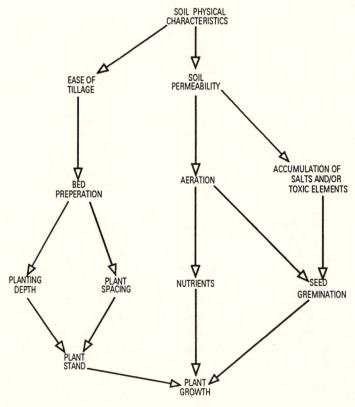

Figure 2-1 Influence of the combination of soil physical characteristics on plant growth.

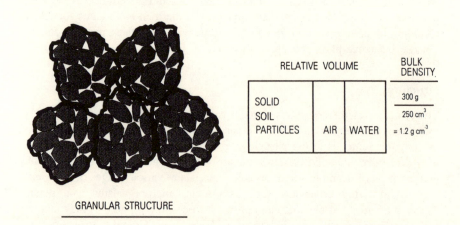

GRANULAR STRUCTURE

Figure 2-2 Schematic of granular soil structure showing aggregation of particles and absorbed water.

aggregate contains small pores with larger pores existing between the granular aggregates. This allows for improved water-holding capacity and permits water movement, root penetration, and aeration. This type of structure generally occurs in medium textured soils as well as in some clay soils.

Prismatic or blocky structure generally occurs in clay loams or in subsoil clay layers as a result of shrinking and swelling of clay minerals during alternate wetting and drying cycles. *Prismatic structure* refers to the existence of elongated chunks of soil between vertical cracks. *Blocky structure* refers to chunks of soil having essentially the same dimensions in all planes. Thin layers of soil particles stacked one on top of the other are referred to as *platy structure*. Soils having this structure are said to have poor physical condition. These thin layers may restrict the movement of water and air through the soil. When no well-defined structural aggregation occurs the soil is referred to as having a *massive structure* (Fig. 2-3). This type of structure can be found in medium and fine textured soils. These soils are often very dense with small and discontinuous pores, which results in poor water and root penetration as well as restricted gaseous exchange.

A type of structure not shown in Figs 2-2 and 2-3 is often referred to as *single-grained*. This structure generally occurs in sandy soils that are low in both organic matter and clay content. Because of the larger particle size of sandy soils they generally have good water penetration and aeration. However, the soils may be compacted to the point where root penetration is inhibited severely. The principal effect of soil structure on plant growth is related to the amount and size of pores in a given volume of soil.

A well-aggregated soil with granular structure has approximately 50 percent of its volume made up of solid soil particles, approximately 25 percent air space, and another 25 percent of water (Fig. 2-2). This means that there is about 50 percent solids and about 50 percent pore space. Since water is held by cohesive and adhesive forces to the surface of soil particles, the water forms a film around individual particles and aggregates and takes up part of the pore space. In a

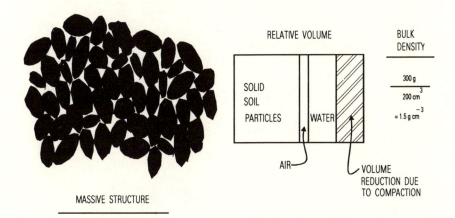

Figure 2-3 Schematic of a soil with massive structure showing the lack of aggregation and reduction in pore volume.

well-aggregated soil the pore space is about equally distributed between air and water. With massive structure (Fig. 2-3), the pore spaces are smaller and water makes up a much larger percentage of the soil pore volume. Hence, the amount of air space present in the soil is very much reduced. According to Baver et al. (1972), the ideal soil should have an equal volume of pore space divided between the large and small pores. Soils with air-filled porosities less than about 10 percent restrict root proliferation.

Due to the larger percentage of small pores in soils with massive structures, aeration and internal drainage are restricted. The effect of soil structure on plant growth was observed in a study (Table 2-1) by Bertrand and Kohnke (1957). The drier soil treatment in the compacted soil resulted in higher yields than the wetter soil treatment, suggesting that poor aeration was the cause of reduced total yields in the wetter soil. The compacted soil had a bulk density of 1.5 (comparable to Fig. 2-3) while the loose soil had a bulk density of 1.2 (comparable to Fig. 2-2). This type of data demonstrates the combined effects of the poor soil structure in compacted soils. Poor structure causes a reduction in the availability of oxygen to the root systems and at the same time impedes root development by requiring the plant to exert greater force in order to extend its root system.

Another well-known effect of inadequate aeration caused by soil compaction is the reduction of nitrate to dinitrogen gas in the soil solution by soil microorganisms. This process is called *denitrification*. It is brought about by the reduced diffusion of oxygen into soil air spaces because of the discontinuous nature of the soil pores. Ardakani et al. (1976) showed that an anoxic environment is created in which denitrification by anaerobic organisms can proceed as soil pores become partly filled with water. While soil structure is an important feature of soils that influences aeration, plant root elongation and water movement, the size of the soil particles is also an important feature affecting many of these same factors.

Texture

The texture of the soil refers to the size distribution of individual mineral particles that make up the bulk of the solid soil matrix. The different particle sizes

Table 2-1.   Effect of moisture, fertility and soil compaction on growth of corn.

Treatment	Weight of tops	Weight of roots	Weight of total plant
Loose soil:	grams/pot	grams/pot	grams/pot
Wet, fertilized	39.4	14.8	54.2
Wet, unfertilized	23.5	10.1	33.7
Dry, fertilized	27.5	9.3	36.8
Dry, unfertilized	20.3	9.3	29.6
Compact soil:			
Wet, fertilized	16.0	6.5	22.5
Wet, unfertilized	17.0	7.7	24.7
Dry, fertilized	20.1	11.3	31.4
Dry, unfertilized	19.3	9.9	29.2

Source: Bertrand and Kohnke, 1957. Copyright © 1957 by the Soil Science Society of America. Reprinted by permission.

are called *soil separates* and are classified into various categories based on their size. There are five different systems of soil separate classification currently used by various groups in the United States (Baver et al., 1972) depending upon whether one is interested in engineering or agricultural features. In each of the classification systems, particles larger than 2 mm, representing gravel and stones, are not to be considered soil particles. The U.S.D.A. system of classification, which is more commonly used by soil scientists in the United States, classifies sand particles in a range from 0.05 to 2 mm, silt particles in a range from 0.002 to 0.05 mm, and clay particles less than 0.002 mm.

The relative proportion of each of these particle sizes in a soil determines its textural class. Figure 2-4 shows the U.S.D.A. soil textural triangle on which the percentages of sand, silt, and clay may be plotted to determine the textural class of the soil. In order to use the textural triangle, one must have a laboratory analysis of the relative percentage of the soil separates. However, with experience it is possible to conduct a field evaluation that approximates the soil texture. The soil is first moistened and then rubbed between the thumb and forefinger while noticing its characteristics. Sandy soils feel gritty when rubbed, silty soils feel slippery but not sticky, and clay soils feel sticky and form a ribbon when extruded between the thumb and forefinger. Laboratory checks of field evaluations provide a mechanism for developing one's expertise in field evaluation.

With a laboratory analysis, the soil can be assigned a textural name based on

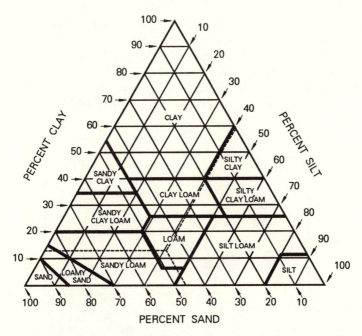

Figure 2-4 Soil textural triangle. The percentages of sand, silt, and clay in any soil may be plotted on this diagram to determine the textural class of that soil. (Wildman and Gowans, 1975.)

the relative percentages of the soil separates. For example, a soil containing 13 percent clay, 41 percent silt, and 46 percent sand would have a loam texture. It is also important to distinguish between soil particle size designation and the mineralogical composition of the soil. Soil textural analysis is based strictly on the size distribution of the soil particles. There is also a good relationship between the presence of highly reactive clay minerals contained in the soil and the clay-sized particles of soils. Clay minerals contribute to the development of soil structure by providing surfaces for binding with water, elements, organic matter, and organisms. Several different types of clay minerals occur in soils. Some expand and contract on wetting and drying, such as montmorillonite. Others do not expand on wetting, such as kaolinite.

The textural differences dictate the type of soil structure which, in turn, determines the air-filled porosity and water holding capacity of the soil. As Black (1957) indicates, there is probably little or no direct effect of soil texture *per se* as with other plant growth factors such as water, oxygen, and mineral nutrient supplies.

The water-holding capacity of soils has been shown to be a function of soil texture. Figure 2-5 shows the change in water content as a function of soil texture and changing soil-water tension. Available water for plant use is the amount of water held in the soil between field capacity (1/3 atm tension) and the permanent wilting point (15 atm tension). Amounts of available water in terms of centimeters of water per 30 cm of soil depth are shown in Table 2-2 for soil textures corresponding to those in Fig. 2-5. In addition, the percentage of water that is available at tensions either above or below 1 atm is listed. This gives an indication of the energy a plant must expend in order to obtain the available soil moisture. The soil water-holding capacity is the primary factor determining the amount and frequency of water application needed to replenish the soil water supply required for plant growth.

Soil texture coupled with soil structure influences the rate of infiltration of water into soils. The range of infiltration rates shown in Table 2-3 is representa-

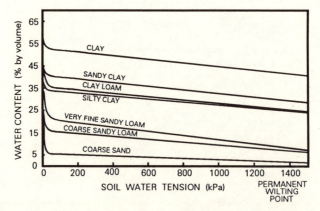

Figure 2-5 Characteristic curves showing volumetric water content for various soil textures as a function of changing soil-water tension. Data from the Soil Characterization Laboratory, University of Florida, Gainesville.

Table 2-2. Available water[a] per 30-cm depth for various soil textures and the relative amount of water removed at soil water tensions above and below 1 atmosphere.

Soil texture	Available water	Percent of available water	
		above 1 atm	below 1 atm
	cm/30cm		
Coarse sand	1.10	95	5
Very fine sand	3.25	78	22
Very fine sandy loam	4.70	64	36
Silt loam	5.30	43	57
Silty clay	6.40	18	82
Clay	7.00	22	78

Source: Lehane and Staple, 1953.

[a] Available water refers to the amount of water expressed in cm/30 cm contained in the soil between soil water tensions representing field capacity and the permanent wilting point.

Table 2-3   Typical ranges in infiltration rates for various soil types used in irrigated agriculture.

Soil type	Infiltration rate[a] ----cm/hr----
Aiken clay loam[b]	9.91
Crown sandy loam[b]	7.24
Palouse silt loam[b]	5.18
Houston black clay[b]	0.30
Oldsmar fine sand (0-10 cm)[c]	18.30
Panoche clay loam (0-30 cm)[d]	0.61

Sources: [b]Free et al., 1940; [c]Selim et al., 1974; [d]Biggar et al., 1976.

[a] Infiltration rate measured at a point in time when it is approaching a very low rate.

tive of different areas of the United States where irrigation is practiced. For soils with high infiltration rates, the rapid movement of water through them contributes to increased potential nitrate leaching below the root zone. For soils with low infiltration rates, there exists a potential for long periods of saturation contributing to anoxic conditions, leading to a relatively higher potential for denitrification.

The low infiltration rate prolongs the time required to replenish the soil water supply by irrigation. During this prolonged irrigation period, there is ample time for denitrification to take place. For example, 30 to 35 hours could be required to replenish the soil water supply for clay soils with poor structure when 50 percent of the available water supply has been removed by plant growth. Soils of this type require special consideration when designing irrigation application systems in order to minimize denitrification losses during irrigation events. Because of the low infiltration rate, the water-holding capacity, and the increased potential for denitrification, the potential for leaching losses below the root zone is limited for this soil type.

Coarse textured soils have higher infiltration rates and are more susceptible to nitrate leaching below the root zone. This occurs for two reasons. First, the soils characteristically have very low water-holding capacity, as indicated in Fig. 2-5 and Table 2-2. This means that the water content of the soil must be replenished more frequently because plants deplete the reservoir much more rapidly. Second, because of the high infiltration rate, the amount of water applied to the soil is very difficult to control. This is especially true where surface irrigation methods are being utilized. Irrigation systems designed to apply water at lower rates or in smaller amounts are less likely to cause nitrate leaching losses below the root zone. Because of the highly porous nature of sandy soils, they are less susceptible to denitrification losses. However, there is mounting evidence to indicate that microsites (micropores) can still exist within the soil profile, where anoxic conditions may exist that allow denitrification to occur in what is normally considered an aerobic soil (Parkin, 1987; Staley et al., 1990; Christensen et al., 1990).

In a study on a Hanford sandy loam, Broadbent and Carlton (1976) showed that between 16 and 25 percent of the added nitrogen over a 2-year period was lost, largely through denitrification. During this experiment, the soil water supply was replenished by overhead sprinklers. Relatively little of the fertilizer nitrogen had been leached below the sampling depth of 15 ft (450 cm) during the study period. Similar information was found by Rolston et al. (1977) for a Yolo loam where soil-water pressure heads of -15 and -70 cm were maintained nearly constant with depth in order to evaluate the effect of different soil-water contents or *in situ* denitrification. They were able to measure considerable oxygen in the zones where denitrification was occurring and therefore concluded that denitrification was occurring in anoxic sites or pockets within the soil. Such conditions probably occurred due to high microbial activity and low oxygen diffusion rate.

Uniformity

Even though soils may be mapped as a single soil type, for which the texture is assumed to be uniform, differences can and do exist in the field. It is rare that a soil is uniform in structure and texture at the surface and with depth.

Soil uniformity has a definite impact on the infiltration of water. First, consider the case of soil that is not uniform with respect to surface texture within a management unit. In large management units, one could generally expect to find greater differences with respect to surface soil texture than for smaller management units. Frequently, the differences in uniformity at the surface are such that it is not feasible to perform different cultural practices for each of the nonuniform areas within the management unit. A diagram of the type of surface soil nonuniformity that can occur within a management unit is shown in Fig. 2-6. This example points out the necessity of knowing the characteristics of a soil on a site-specific basis. In this example it is not economically practical to arrange management units of lesser size or different shape to contain soils uni-

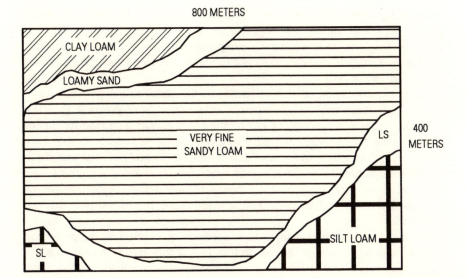

Figure 2-6 A diagram of the nonuniformity of soil textures that may occur within management units.

formly of the same texture. The reason for this is the size of equipment used and the cultural practices that must be performed in order to grow a crop. Farm managers must optimize the resources (machinery, labor, management, capital) available to them to produce a crop with present-day technology. One difficulty of having to manage a management unit with such variation in soil texture is that different amounts of water percolate through each soil type as a function of its water-holding capacity and infiltration rate.

A practical problem in water management arises when one attempts to apply adequate water for plants in the field. Because of a lower water-holding capacity, a loamy sand would have its water supply exhausted first. Plants growing on this portion of the management would then exhibit water stress; whereas plants growing on the finer textured soils would not be exhibiting stress. If water were applied to accommodate the plants grown in the loamy sand, then excess water would be applied to the other soil types, leading to increased leaching and denitrification potential. Conversely, if water is withheld from the field until plants on the finer textured soils require irrigation, then the plants growing in the loamy sands would exhibit severe water stress and yield reductions would occur.

The general practice is to apply water to the field according to the water depletion in the soils representing the major portion of the management unit. In doing so, one implicitly recognizes that less than optimum soil moisture conditions will exist for either the finer or coarser textured soils within the unit. In assessing the overall non-point source pollution potential for a management unit, one must evaluate the potential of nitrogen leaching from overirrigation of coarser textured soils against denitrification from overirrigation of finer textured soils.

Where extreme variation exists, the disparity in infiltration rate and water-holding capacity of the soils in the management unit with their subsequent impact on denitrification, leaching, and surface runoff may be partially overcome by irrigating more frequently using small amounts of water each time. This type of irrigation practice is most easily accomplished by use of a sprinkler or drip irrigation system.

A second dimension of soil uniformity is the stratification with respect to depth. Such strata may include changes in texture or structure, or both. They normally occur as a function of soil-forming factors, although modification of the soil structure may result from tillage operations. In older soils that have developed in place, one frequently finds increasing amounts of clay with depth in the soil profile. In recent alluvium (soils deposited by water movement), different strata may develop as a function of how rapidly water has moved over the landscape surface causing differential deposition of soil particles, and parent material.

Textural and structural changes have a tremendous impact on the movement of water and roots through the soil profile. This in turn affects the rate of nitrogen loss in the soils. Differences in infiltration rate as a function of soil depth can be seen in Fig. 2-7. The differences in intake rate with various depths in this case were attributed to soil compaction. For the Wyo silt loam and the Cajon sandy loam, the surface had apparently been compacted and is causing the relatively low infiltration rate. The Hanford fine sandy loam with a relatively low intake rate throughout the soil profile had only a slightly reduced intake rate at the surface.

Relatively low intake rates create a situation where water must be applied for long periods of time in order to insure adequate recharge of the partially dried soil profile. It is during relatively long periods of application that aeration is restricted. This interferes with plant root development and creates a potential for increased denitrification. In a compacted soil there is impedance of root development, resulting in a reduction in the extensiveness and density of the root system which, in turn, influences nitrogen uptake.

Data presented in Fig. 2-8 illustrate the effects of layered soils on the water content distribution and rate of movement of water through the profile. The resulting high water content and its influence on aeration has an adverse effect on plant root development and increases the denitrification potential. If restricting layers occur relatively near the soil surface, the plant root system can be severely restricted, which interferes with plant water uptake and nitrogen utilization.

Depth

Soil depth may be limited by the existence of strata within the soil profile that restrict a plant's root development from reaching its genetic potential. Restrictive layers are generally the result of parent material, clay pan, hard pan, or a relatively thick stratum of soil having a distinctly different texture from the layers above or below. Terms commonly used to describe soil depth are shown in Table 2-4.

Because of the relative impermeability of restrictive layers to roots and water,

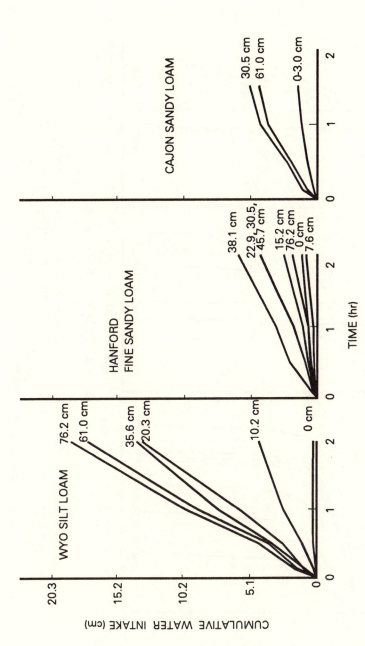

Figure 2-7 Cumulative water intake as a function of different soil depths for three soils (Cajon sandy loam (mixed, thermic type torripsamments), Hanford fine sandy loam (coarse-loamy, mixed, nonacid, thermic typic Xerorthents) and Wyo silt loam soils of uniform soil texture with depth). (Wildman, 1969.)

21

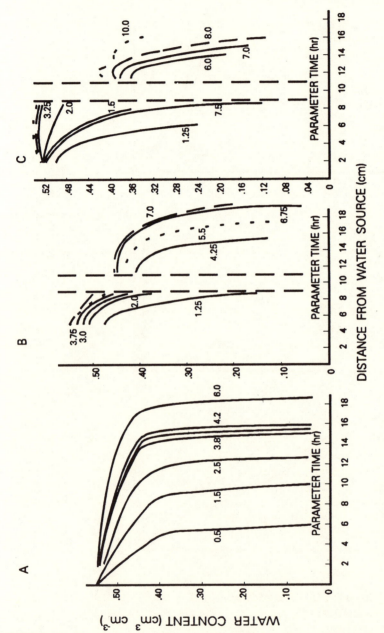

Figure 2-8 Influence of a uniform (A) and a stratified (B,C) soil on water movement through a Salkum silty clay loam (clayey, kaolinitic, mesic typic Hapludolls). ((A,C) Ferguson and Gardner, 1962. Copyright © 1962 by the Soil Science Society of America. Reprinted by permission; (B) Ferguson, 1959.)

Table 2-4. Terms used to describe soil depths.

Terms	----------------Depth range, cm----------------	
	Storie[*]	USDA[b]
Very shallow	less than 30	less than 25
Shallow	30 to 61	25 to 51
Moderately deep	61 to 91	51 to 91
Deep	91 to 122	more than 91
Very deep	more than 122	more than 152

Source: Wildman and Gowans, 1975.

[*] Storie, W. Earle. Soil Series of California. University of California, Berkeley. 1953. Associated Student Store.

[b] USDA. Soil Survey Manual. Bureau of Plant Industry, Soils, and Agricultural Engineering. 1951.

they have a significant impact on water movement, plant development and soil nitrogen pools. The effect of a clay pan at 12 inches on the intake rate of a Huerhuero sandy loam is shown in Fig. 2-9. The accumulation of water above the strata creates a condition referred to as perched water table. The anoxic conditions created by the perched water table inhibit root development and create a condition conducive to denitrification losses. Inhibited root development results in limited nutrient absorption and a yield reduction. The extent of denitrification that occurs will determine the concentration of nitrate that may be leached below the restricting soil layer.

SOIL FERTILITY EVALUATION

Soil fertility is the result of a combination of physical and chemical soil characteristics, allowing plants to obtain a continual and adequate supply of nutrients, a readily replenishable reservoir of water, and a physical matrix that supports the plant, does not impede root development, and provides good aeration. A combination of these factors influences the growth and yield of a given crop. Until relatively recently, there has been no systematic attempt to delineate soil fertility in relation to soil chemical and physical characteristics that influence the nutrient supply to the plant and subsequent plant growth.

Fertility Classification

A system of classifying soils with respect to their fertility capability was proposed by Buol (1972) and expanded upon by Buol et al. (1975) and later by Sanchez et al. (1982) (Table 2-5). The proposed system utilizes existing soils data from soil survey reports, or on-site examination of areas where surveys are not available. The information is used to group soils into various classes for the purpose of soil fertility interpretation. The system is not meant to replace soil

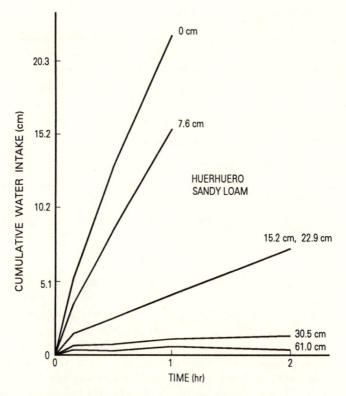

Figure 2-9 Cumulative water intake for a Huerhuero sandy loam soil with a clay pan at 30.5 cm below the soil surface. (Wildman, 1969.)

testing, which is necessary to monitor annual changes in soil fertility levels due to management practices. Its purpose is to provide the soil fertility specialist with somewhat uniform soil groups within which they can feel comfortable about extrapolating soil test information.

CHEMICAL CHARACTERISTICS

The chemical characteristics of soils have a marked influence on plant development and nitrogen pools within the soil profile. Clay mineral content of soils plays an important role in regulation of nutrient supply to plants. One of the most important of these characteristics is the cation exchange capacity (CEC). The crystalline nature of clay minerals and the type of bonding associated with the elements that make up these inorganic polymers results in a net negative electrostatic charge at their surfaces. The negative charge is neutralized by the association of positively charged ions (cations) from the soil solution.

Different clay minerals, because of the manner and degree to which the charge is generated, have different capacities for adsorbing cations on their surfaces and exchanging with other cations in soil solution. In Table 2-6 are shown three

Table 2-5.   Fertility-capability classification system.

The system consists of three categorical levels: type (topsoil texture), substrata type (subsoil texture), and 15 modifiers, including several changes from the original version (Buol et al, 1975) making the following , in effect the second approximation. The classes within each categorical level are defined below. Class designations from three categorical levels are combined to form a FCC-unit.

TYPE:
Texture of plow-layer or 20 cm (8") depth, whichever is shallower:
S = Sandy top soils: loamy sands and sands (by USDA definition).
L = Loamy top soils: <35% clay but not loamy sand or sand.
C = Clayey top soils: >35% clay
O = Organic soils: >30% O.M. to a depth of 50 cm or more.

SUBSTRATA TYPE:
Used only if there is a marked textural change from the surface or hard root-restricting layer is encountered within 50 cm:
S = Sandy subsoil: texture as in TYPE.
L = Loamy subsoil: texture as in TYPE.
C = Clay subsoil: texture as in TYPE.
R = Rock or other hard root-restricting layer.

MODIFIERS:
Where more than one criterion is listed for each modifier, only one needs to be met. The criterion listed first is the most desirable one and should be used if data are available. Subsequent criteria are presented for where data are limited.

g = (Gley): Mottles $\leq$ 2 chroma within 60 cm of surface and below all A horizons or saturated with H_2O for > 60 days in most years;

d = (dry): Ustic, aridic, or xeric moisture regimes (subsoil dry >90 consecutive days per year within 20-60 cm depth);

e = (low cation exchange capacity): applies only to plow layer or surface 20 cm, whichever is shallower: CEC < 4 meq/100 g soil by Σ bases + KCl-extractable Al (effective CEC), or CEC < 7 meq/100 g soil by Σ cations at pH 7, or CEC < 10 meq/100 g soil by Σ cations + Al + H at Ph 8.2;

a = (aluminum-toxicity): > 60% Al saturation of CEC by (Σ bases and unbufferred Al) within 50 cm of the soil surface, or > 67% aluminum saturation of CEC by Σ cations at Ph 7 within 50 cm of the soil surface, or > 86% aluminum saturation of CEC by Σ cations at pH 8.2 within 50 cm of the soil surface, or pH > 5.0 in 1:1 H_2O within 50 cm, except in organic soils where pH must be less than 4.7.

h = (acid): 10-60% Al-saturation of CEC within 50 cm of soil surface or pH in 1:1 H_2O between 5.0 and 6.0.

i = (high P-fixation): % free Fe_2O_3/ % clay > 0.15, or more than 35% clay, or hues of 7.5 YR or redder and granular structure. This modifier is used only in clay (C) types: it applies only to plow-layer or surface 20 cm of soil, whichever is shallower;

x = (X-ray amorphous): pH > 10 in $1N$ NaF, or positive to field NaF test, or other indirect evidences of allophane dominance in clay fraction;

v = (Vertisol): Very sticky plastic clay > 35% clay and > 50% of 2:1 expanding clays, severe topsoil shrinking and swelling;

k = (low K reserves): < 10% weatherable minerals in silt and sand fraction within 50 cm of soil surface, or exchangeable K < 0.20 meq/100 g, or K < 2% of Σ of bases; if Σ of bases < 10 meq/100 g;

Continued ---.

TABLE 2-5 (continued)

---

b = (basic reaction): Free CaCO$_3$ within 50 cm of soil surface (effervescence with HCl); or pH > 7.3;

s = (salinity) ≥ 4 mmhos/cm of electrical conductivity of saturated extract at 25°C within 1 m of the soil surface;

n = (natric): ≥ 15% Na saturation of CEC within 50 cm of the soil surface;

c = (cat clay): pH in 1:1 H$_2$O is < 3.5 after drying and Jarosite mottles with hues 2.5Y of yellower and chromas 6 or more are present within 60 cm of the soil surface;

' = (gravel): a prime (') denotes 15-35% gravel or coarser (>2mm) particles by volume to any type or substrata type texture (example: S'L = gravelly sand over loamy; SL' = sandy over gravelly loam); two prime marks ('') denote more than 35% gravel or coarser particles (>2 mm) by volume in any type or substrata type (example LC'' = loamy over clayey skeletal; L'C'' = gravelly loam over clayey skeletal);

% = (slope): where it is desirable to show slope with the FCC, the slope range percentage can be placed in parentheses after the last condition modifier (example: Sb (1-6%) = uniformly sandy soil, calcareous in reaction, 1-6% slope).

The soils are classified by determining if the characteristic is present or not. Most of the quantitative limits are criteria present in Soil Taxonomy (USDA, Soil Survey Staff, 1975)

---

clay minerals commonly found in soils with a description of their crystalline structure and characteristics.

Adsorbed ions do not adhere tightly to the surface of clay minerals but are distributed in the soil solution surrounding the mineral particles. This permits the exchange of a cation from the soil solution with one more closely associated with the surface of the clay mineral. As seen in Fig. 2-10, the cations in soil solution tend to concentrate near the negatively charged surface of the clay mineral due to the electrostatic attraction. They become less concentrated as their distance from the surface increases. The greater the distance from the surface of the charged clay mineral, the more likely that negatively charged ions (anions) will be interspersed with the cations. Nitrogen in the form of ammonium-N (NH_4^+-N) in soil solution exists as a positively charged ion. It is attracted to the surface of clay minerals, where it is absorbed in the manner indicated. This phenomenon prevents the leaching of ammonium-N through the soil profile. It is only after ammonium-N is transformed to nitrate anion (NO_3^--N), which is repelled by the negatively charged soil particles, that nitrogen is susceptible to leaching.

The CEC of soils influences the nitrogen pool in soils in another manner by providing adsorption sites for ammonium-N, thereby controlling the chemical change to ammonia-N (NH_3-N), which volatilizes from the soil surface. The

Table 2-6. Selected characteristics of three important clay minerals found in soils.

Clay mineral	Schematic of crystalline structure	Source of electrostatic charges	Relative cation exchange capacity	Shrink-swell properties	Field evidence of clay minerals
Kaolinite	1 Silica : 1 Alumina Sheet	Ionization of hydrogen from hydroxyl groups on the surface of the crystal layer and at the broken bonds at the edges of the fractured crystal.	3-15 milliequivalents per 100 g of oven-dry soil.	Does not shrink and swell with drying and wetting because of tendency of surface bonding to occur between crystal layers.	Drying does not cause soil to crack.
Illite	2 Silica : 1 Alumina Sheet	Hydroxyl groups and substitution of certain elements in crystal lattice, e.g. Iron (Fe) and Magnesium (Mg) in the Alumina sheet and aluminum (Al) in the silica sheet.	10-40 milliequivalents per 100 g of oven dry soil.	Does not shrink or swell upon drying and wetting because of strong surface charges and existence of Potassium (K) between crystal layers assisting in holding layers together.	Drying does not cause soil to crack.
Montmorillinite	2 Silica : 1 Alumina Sheet	Hydroxyl groups and substitution of certain elements in the crystal lattice, e.g. Magnesium (Mg) in the Alumina sheet.	50-80 milliequivalents per 100 g of oven dry soil.	Does shrink and swell upon drying and wetting.	Soil cracks upon drying and swells so that cracks disappear upon wetting.

DISTANCE FROM MINERAL SOURCE

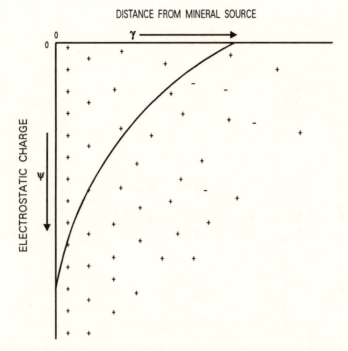

Figure 2-10   Ion and potential distribution near a charged surface. (Nielsen et al., 1972. Copyright © 1972 by the American Society of Agronomy and the Soil Science Society of America. Reprinted by permission.)

relationship between the total ammonium-N losses and depth of nitrogen place-ment in soils with different CEC values is shown in Fig. 2-11. The Houston black clay and Sharky silty clay loam used in this study had CEC values of 58 and 22 meq/100 g, respectively. The pure sand had a CEC of 0. These data demonstrate the relative effect of CEC on volatilization losses. They are not meant to represent the actual magnitude of the losses that will occur from any given soil.

Ammonium may also become fixed by clay minerals. According to Gasser (1965), two main theories have been postulated for this process. One is that ammonium-N enters into the lattices of the mineral, and another is that ammonium-N is trapped in interlayer exchange positions in much the same man-ner as potassium. This phenomenon is described for the clay mineral illite in Table 2-6. In either case, the ammonium-N is not readily available for exchange with cations in soil solutions, and consequently, is referred to as having become fixed. The magnitude of fixation may range from negligible to over 500 ppm as shown in laboratory experiments by Allison et al. (1953). While this phenom-enon is known to exist, it is not common in irrigated soils that are relatively low in illite mineral content. Consequently, it has to be diagnosed on a site-specific basis.

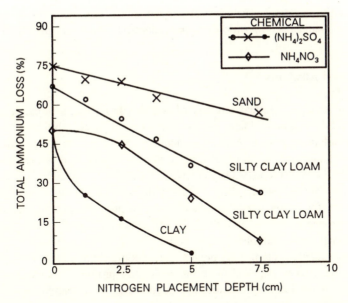

Figure 2-11 Influence of soil type and placement depth on NH_4^+-N losses. Ammonium sulfate and ammonium nitrate were applied at 550 kg/ha NH_4^+-N at 30°C. (Fenn and Kissel, 1976. Copyright © 1976 by the Soil Science Society of America. Reprinted by permission.)

Soil Reaction

Another chemical characteristic of soil influencing plant growth and the nitrogen pool is the soil reaction (pH), which is indicative of the acidity or alkalinity of the soil. Optimum soil pH for most plants occurs in the range of 6 to 8 (moderately acidic to moderately alkaline). Depending upon the species, plants may grow relatively well either below or above the optimum level indicated. A reduction in yield associated with sub-optimal pH range leads to a reduction in the amount of nitrogen required to obtain the lower yield.

In addition to its direct effect on plant growth, there is also an effect of pH on atmospheric nitrogen fixation. Black (1957) indicated that certain cultures of *Azotobacter* can fix nitrogen at a pH of 3.1. *Azotobacter* is an aerobic, free living organism. An anaerobic organism capable of fixing nitrogen is *Clostridium*. Although it can be found in soils with pH as low as 5, it is generally found in soils having a pH greater than six (Black, 1957). Symbiotic fixation in agricultural soils is generally a result of *Rhizobium* bacteria which creates root nodules on leguminous plants. In a review article by Munns (1977), it was indicated that at pH levels below 5, symbiotic fixation in alfalfa is very much reduced. It was also shown that *Rhizobium* species could be found that were more tolerant of low soil pH. In one case he selected an organism that continued to fix nitrogen in association with subclovers to a pH of 4.6.

Soil pH also has an effect on organic matter decomposition. It is recognized that soil fungi become more active at lower pH values than bacteria. However, fungi usually decompose materials at a much slower rate. As a result, the rate of organic matter decomposition may be reduced at lower soil pH levels.

Salinity and Alkalinity

Salinity is another chemical property of soil that has a major impact on plant growth and development. Plant species have different genetic abilities to tolerate salt accumulations in soils. Numerous publications have been written that give relative salt tolerance of crops. Perhaps the most widely used is the Food and Agriculture Organization (FAO) publication entitled *Water Quality For Agriculture* (Ayers and Westcot, 1985). Yield reduction by salinity decreases the requirement for nitrogen by the plant. Consequently, the potential exists for greater nitrogen carryover and, thus, the nitrogen pollution potential increases wherever salt has accumulated in the soils.

Salt accumulations are usually prevented by adding slightly more water during an irrigation than is required to recharge the soil profile. This excess water, referred to as the leaching fraction, is recognized as a reasonable use of water by state water resources agencies. Since salinity management is an essential feature of irrigated agriculture and is accomplished by leaching of soil profile during an irrigation, it follows that nitrate-N has the potential to be leached in the process. Consequently, the timing of nitrogen applications should take into account this process in order to provide the greatest efficiency of nitrogen use.

A related chemical characteristic of soils that has a major influence on structure is the exchangeable sodium percentage (ESP). A rather arbitrary value of 15 percent has been used to delineate between affected and slightly affected soils. As ESP increases, the amount of dispersion of soil aggregates increases, creating a massive structure. The impact on plant development and denitrification is the same as discussed earlier for the massive soil structure. Some plant species have been shown to be sensitive to excessive amounts of sodium in soils as a specific ion effect.

Specific Ion Toxicities

Some nonnutrient elements as well as some nutrients required by plants can become toxic if present in soils in excessive amount. The nutrient boron is a classic example. Boron tolerances for various plants are given in *Water Quality For Agriculture* (Ayers and Wescot, 1985). The importance of these specific ion effects, with respect to nitrogen in the environment, is primarily related to the reduced nitrogen requirement resulting from lowered yields. The result can be higher nitrogen carryover and a greater potential for leaching of nitrate. The specific ion toxicities most often observed are those associated with sodium, chloride, boron, manganese, copper, and on occasion, zinc and molybdenum. Other toxic elements are also of concern, especially those related to sewage sludge disposal on agricultural soils (e.g. cadmium and lead).

Organic Matter

The organic matter content of soils is a function of additions of organic waste, fertility, and climatic factors. White et al. (1976) showed increases in organic matter contents over an 8-year period for different types of pasture in North Dakota. The soils had an organic matter content on the order of 3.5 percent. However, in most irrigated soils, especially in arid areas where temperatures seldom drop below freezing, soil organic matter contents are much lower. They generally range from about 0.25 percent to 2 percent. Even with changes in the amount of crop residue or animal waste applied to such soils, the change in soil organic matter content is generally quite small. In the short term, the change is almost undetectable. For example, even in the study by White et al. (1976), they only observed about a 0.02 percent increase in soil organic matter per year for the 8-year period. This occurred under conditions where one would expect organic matter contents to accumulate at a higher level than in areas where the soil is never frozen. Decomposition proceeds on a year-round basis in areas where soils do not freeze.

Organic matter has traditionally been considered the relatively stable end product of the decomposition of crop and animal wastes in soils. Common usage of the term organic matter refers to any organic material in soils, regardless of its state of decomposition. It is important to distinguish between the stable and unstable organic matter in the soil. The relatively undecomposed organic residue, even with high carbon to nitrogen ratios (C : N), may release more N to the soil environment than stable soil organic matter with a much lower C : N ratio (10 : 1). So-called soil humus is not readily decomposed by soil microorganisms.

Carbon : nitrogen ratios of plant residue may range from over 100 : 1 for materials such as rice straw, to less than 20 : 1 for most legumes. At C : N ratios of about 20 : 1, there is more than enough nitrogen present to meet growth requirement of microorganisms during decomposition of the material. Consequently, nitrogen is released to the surrounding soil environment. At C : N ratios over 40 : 1, adequate nitrogen for microbial growth is not present in the organic material to produce rapid decomposition; therefore, the organisms utilize nitrogen from the soil environment. In the presence of growing plants, this generally induces a nitrogen deficiency. By virtue of their large numbers in the soil environment, microorganisms have a competitive advantage over plant roots in utilizing inorganic nitrogen in the soil solution.

The effect of addition of various amounts of straw on nitrogen content of rice plants is shown in Table 2-7 (Rao and Mikkelsen, 1976). These data contradict the commonly held misconception that nitrogen additions are required to permit decomposition of straw without affecting the nitrogen nutrition of the plant. The controlling factor is the timing of planting after incorporation of the straw. With no incubation of the soil–straw mixture prior to planting, a significant reduction in plant nitrogen content occurred with increasing straw additions. These investigators found that incubation of soil–straw mixture for 15 to 30 days prior to planting of rice seedlings eliminated nitrogen deficiencies. This

Table 2-7.   The effect of rice straw added to soils on the nitrogen content of rice
              plant parts.

Number of days incubated	Percent[a] of straw added	Percent nitrogen in	
		Leaf	Sheath
	--------%--------	-------------- % --------------	
0	0	3.73	2.26
0	0.25	3.41	1.29
0	0.50	1.89	1.15
15	0	3.38	1.72
15	0.25	3.19	1.72
15	0.50	2.87	1.83
30	0	3.51	1.79
30	0.25	3.34	2.05
30	0.50	3.12	1.84

Source: Rao and Mikkelsen, 1976. Copyright © 1976 by the American Society of
Agronomy. Reprinted by permission.

[a] Straw was added as percent of soil by weight.

information corroborates results obtained by numerous other investigators showing that allowing sufficient time for decomposition, even for materials having
relatively high C : N ratios, will generally result in a net release of nitrogen to
the soil environment. Similar information was reported by Sain and Broadbent
(1977). They found an increased rate of decomposition for a period of about 3
weeks after nitrogen had been added to rice straw as opposed to the case where
no nitrogen had been added. After 60 days, the differences between the amount
of decomposition were slight. At the end of a 120-day incubation period, the
total amount of carbon lost was the same whether nitrogen had been added or
not. Rice straw decomposition may be accelerated by additions of nitrogen,
though it is not recommended as a management practice unless another crop is
to be planted soon after residue incorporation.

One of the other benefits of soil organic matter is its influence on soil structure. Organic by-products from the active decomposition of animal and crop
residues are the materials that cement individual soil particles and assist in the
formation of soil aggregates. The principal metabolic by-products of decomposition responsible for the formation of soil aggregates are slime, mucilages, and
polysaccharides (Broadbent, 1953).

SOIL TESTING FOR EVALUATING NITROGEN STATUS OF SOILS

The value of soil tests in determining the nitrogen status of soils and for predicting the response to added nitrogen fertilizer is the subject of continuing
debate among soil scientists. Most of the controversy centers around the
reliability of the soil test ranges, from the use of a soil test value as an estimate
of the absolute amount of nitrogen available for a crop to the concept that the

soil test value is roughly a guide for estimating nitrogen fertilizer needs. It does not substitute for other information that one may use for determining nitrogen fertilizer needs, but it does enhance the probability of making a correct decision as to amount of fertilizer nitrogen needed.

Because of the transitory nature of nitrogen in soils, variability exists with regard to available forms of nitrogen in the root zone. This reduces the reliability of the soil test for predicting nitrogen requirements. A systematic approach to the collection of samples was proposed by James et al. (1967) in order to eliminate some of the variability known to occur. Nelson et al. (1965), in recognition of the variability in nitrate-N concentrations that exists in soils as a function of irrigation methods, suggested that three cores of soils be taken across a distance equal to one half the ridge or bed spacing for fields where furrows or rills were utilized to manage water. For example, if rows were 36 inches (91 cm) apart, take the first sample any place in the field; the next sample would be taken 9 in (23 cm) from the first at right angles to the row, and the third sample 18 in (46 cm) from the first at right angles to the row. Sampling is in the bed or in the furrow. They presumed that a composite of the three cores provided a method of overriding or averaging out the known variability that exists in the furrow-irrigated fields. For sprinkler-irrigated fields, they took only one sample at each sampling point.

Studies by Reuss et al. (1977) and Ludwick et al. (1977) have examined nitrate concentration with respect to the horizontal and vertical distribution in soils. Similarly, Roberts et al. (1976) evaluated the nitrate content in the 0 to 60 cm depth with respect to sugar beet response to added nitrogen. The predictive capability of the soil test was not enhanced by including the nitrate below the 60 cm depth in the soil test index.

The precision with which one can estimate the nitrate-N level in a field is another interesting aspect of the sampling variability. Reuss et al. (1977) found that 82 cores per field were required for a 90 percent confidence interval of plus or minus 15 percent. In analyzing the data for nitrate concentration in the soil samples, the data were found to be highly skewed. By utilizing a logarithmic transformation, the skewness of the data was markedly reduced. This is consistent with findings by Biggar and Nielsen (1976) and Nielsen et al. (1974). They showed that the spatial variability of the leaching characteristics and other soil-water properties measured in the field are log-normally distributed. Since nitrate moves with water, the concentration of nitrate in soil profiles is distributed in much the same manner.

The spatial variability of nitrate concentration in a field presents a serious limitation to using soil sampling for nitrogen as a basis for making precise recommendations. Reuss et al. (1977) concluded that it does not prevent the use of the soil test as a fertilization guide.

Numerous investigators have had remarkable success in using soil tests for nitrogen to make fertilizer recommendations. A majority of the 17 western states, representing much of irrigated agriculture, utilize soil tests for nitrogen in making fertilizer recommendations on a large variety of crops. One approach has been to use a soil test nitrogen index, as proposed by James et al. (1967), James (1971), and Roberts et al. (1976), as a measure of soil nitrogen available to sugar beets. The procedure requires sampling in 1-ft (30-cm) increments to a depth of 6 ft (180 cm) or to the restrictive layer limiting root development, whichever occurs first. Based on these studies, it was found that response of sugar beets to nitrogen fertilizer was highly likely at a soil index of less than 10, whereas it was highly unlikely at a soil index greater than 20.

Studies conducted by Carter et al. (1974, 1975) indicate that mineralizable nitrogen in several Idaho soils constitutes 49 to 81 percent of the total nitrogen available for plant growth where no nitrogen fertilizer was applied. In those instances where mineralized nitrogen makes a major contribution to the soil nitrogen status, it should be taken into account in predicting fertilizer response. However, in studies conducted by Roberts et al. (1976), inclusion of mineralized nitrogen did not enhance the predictive capability of the soil test. Cropping sequence and cultural practices, coupled with soil and climatic differences, would contribute to the differences found by each of these investigators.

Another method of utilizing soil tests for nitrogen as a means of predicting fertilizer response has been to sample the soil profile to a depth of 3 ft (90 cm) and use a linear correlation equation to calculate the yield for sugar beets based on the available soil nitrogen (Hills, 1976).

Several other approaches have been used. In Oklahoma, Baker and Tucker (1973) suggest that nitrogen fertilizer requirement be based upon the potential yield as determined from past yield history. In Arizona, soil test nitrogen is used as a guide to preplant fertilization, and plant analysis is used to determine the additional nitrogen needs. This approach has been used by Gardner et al. (1976) on wheat and by Ray et al. (1964) on cotton. One other approach that has been the use of the combination of nitrate-N plus ammonium-N in a 2-ft (60-cm) soil profile to predict plant response to nitrogen. High correlations were found between soil nitrogen levels and yield of potatoes by Kerbs et al. (1973) and yield of wheat by Whitney (1972).

There is little doubt that soil tests for nitrogen can be a useful guide in helping to determine fertilizer requirements for a large variety of crops under a wide range of climatic and soil conditions. It is equally clear that these current tests can rarely be used to accurately determine the nitrogen available to plants or subject to leaching. The one thing to keep uppermost in mind is that soil test values do not substitute for everything else one knows in relation to the agricultural production system. Since nitrogen is so transitory in soils, it is necessary to have an understanding of interaction between the irrigation system and the soil, the climatic impact on nitrogen availability and transformations, the potential yield of the crop, and a host of other factors. Lest one become discouraged, it should be pointed out that persons familiar with crop production are evaluating such factors and their interactions every day in order to make proper management decisions. Consequently, it seems reasonable to expect that one can utilize the same information to place in perspective a soil test nitrogen value.

For those cropping systems where calibration studies have shown soil testing to be effective, the tests can be useful guides in predicting responses to nitrogen fertilizer. Even in areas and for crops where calibrated soils tests have not been developed, the general guides used in other areas may be useful. However, they must be used with caution and a great deal of knowledge about the prevailing conditions.

3

Crops

The intent of this chapter is to examine principles that govern crop responses to nitrogen fertilization. An understanding of them should increase the efficiency of nitrogen fertilizer use by improving the prediction of nitrogen fertilizer requirements. The result should be a reduction in the nitrogen pollution potential.

Although the principles governing crop behavior with respect to nitrogen fertilization are similar, different plant species are unique in their patterns of nitrogen demand during a growing season, removal of nitrogen in the harvested portion of the plant, storage of nitrogen within the plant, and response to different levels of soil nitrogen. Frequently, one or more of these factors can vary between cultivars within the species. Because of these differences, it is necessary to become familiar with the growth characteristics and nutritional needs of each plant species or variety to properly employ practices to utilize nitrogen in the most efficient manner.

NITROGEN IN PLANTS

Metabolic Constituents

Nitrogen in one of its various forms is involved in nearly every plant metabolic process. It is found in heterocyclic compounds that are integral parts of genetic materials and plant growth hormones. The principal plant nitrogen pools are proteins, which are made up by combination of amino acids and amides.

A nitrogen compound unique to plants is chlorophyll. Chorophyll is responsible for the transformation of light energy from the sun into compounds having high-energy bonds. Animal life is totally dependent upon the photosynthetic process to provide sugars, fats, proteins, essential amino acids and vitamins. Plants also provide inorganic nutrients that are usable by animals.

Changes in Concentration

Specific nitrogenous compounds have been shown to be extremely important in relation to the nitrogen status of the plant. The ammonium and nitrate fractions are the predominant inorganic forms of nitrogen in the plant. Alcohol-insoluble nitrogen (generally considered to be protein) is the predominant organic fraction of nitrogen within the plant. As pointed out by Rauschkolb (1968) on cotton and Rauschkolb et al. (1974a,b) on corn and sorghum, the total nitrogen content in various tissues is highly indicative of the plant nutritional status. Nevertheless, one must be aware that the nitrogen pools within the plant are in a dynamic equilibrium and that the concentration of a particular constituent is only indicative of the nitrogen distribution in a plant at one point in time.

Frequently, the nitrate pool in certain tissues of the plant is measured to determine the nitrogen status. Plant parts that are predominantly xylem tissues are used since nitrate-N concentration within these tissues is highly sensitive to soil-nitrogen supply.

It is very important to identify and select plant parts for analysis that experience the greatest shifts in a particular nitrogen pool as a result of differences in nitrogen supply to the plant. In general, storage tissues (such as leaf tissue) are most desirable for determining total nitrogen, whereas conductive tissues (such as petioles and stems) are most desirable for determining plant nitrate-N levels.

The usefulness of the nitrate and total nitrogen pools in various plant tissues in evaluating the nutritional status of plants is pointed out in Figs 3-1 and 3-2, and Tables 3-1 and 3-2. The total Kjeldahl nitrogen concentration in the top and roots of 45-day-old cotton plants grown in culture solution continued to decrease the longer the plants grew without nitrogen (Figs 3-1 and 3-2). The plant protein fraction served as a sink for stored nitrogen that was mobilized and redistributed through the plant in order to partially meet the nitrogen demand of rapidly growing plant tissue in a deficient condition. The ability to degrade and translocate part of the nitrogen protein pool is what leads plant nutritionists to call nitrogen a mobile nutrient within plant tissue. It also dictates the type of deficiency symptoms one observes, as will be discussed in a later section. Figures 3-1 and 3-2 also show the responsiveness of this nitrogen pool to fertilizer nitrogen sources as a function of the level of nitrogen deficiency. When the cotton plant has only experienced adequate or mildly deficient nitrogen conditions, there does not appear to be a preference for either of the nitrogen sources. However, as cotton seedlings become severely deficient, they show a preference for the nitrate-N.

In studies by Rauschkolb et al. (1974a,b) the total nitrogen content of whole leaf tissue was satisfactory for determining the nitrogen status of both corn and sorghum (Tables 3-1 and 3-2). As plants aged, the nitrogen concentration in the tissue decreased regardless of the level of nitrogen fertilization. Nitrate concentrations of different plant tissues were also studied as a function of rate of nitrogen application. The basal stem tissue showed a more gradual decrease in nitrate-N concentration than did the midrib. The midrib tissue is not very useful as a means of evaluating the nitrogen status of these crops because of the rapid shift in nitrate-N concentration over short time periods. Where different tissue analyses have comparable capability of evaluating the nitrogen status at the plant, the procedure more readily adapted to routine laboratory analysis is preferred.

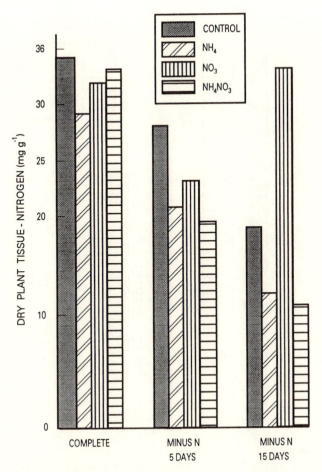

Figure 3-1 Total Kjeldahl-N fraction in cotton tops for different nitrogen application regimes and sources. (Rauschkolb, 1968.)

In tissue testing it is very critical to select tissue for analysis at the same physiological stage of development in order to permit evaluation of nutritional status over time. As already pointed out for cotton, corn, or sorghum, the nitrogen concentration in the tissues sampled changes with time. As the plant matures, various stored forms of nitrogen are utilized for fruit and seed development. Sugarbeets are a classic example where low nitrogen levels are essential near the end of the growing season in order to achieve the desired combination of root yield and sugar percentage (Fig. 3-3).

Data in Fig. 3-4 show the relationship between the petiole nitrate-N concentration and the yield of potatoes associated with different levels of nitrogen application for a nitrogen-deficient soil. The data show that some minimum level of nitrate-N must be maintained in the petiole to achieve maximum yield. Figure 3-5 shows the relationship between petiole nitrate-N concentration and potato yield as a function of applied nitrogen. Soil tests indicated a relatively high nitrate-N level. With each successively higher rate of applied nitrogen,

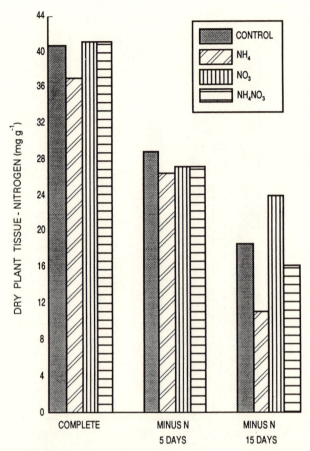

Figure 3-2 Total Kjeldahl-N fraction in cotton roots for different nitrogen application regimens and sources. (Rauschkolb, 1968.)

there was a decrease in the potato yields. Consequently, Jones and Painter (1974)
proposed that petiole nitrate-N levels of potatoes should follow that indicated
in the adequate zone in Fig. 3-6 in order to achieve maximum yield. Either
inadequate or excessive nitrate-N levels may cause a decrease in yield. Similar
curves have been developed by other investigators for numerous other crops.

   In perennial deciduous plants the amino acid nitrogen pool has also been
shown to reflect the nitrogen status of the plant. This has generally been associated with changes in concentration of a particular amino acid or amide in the
sampled tissue. Because of qualitative and quantitative changes in these compounds at various physiological stages of development, it has been very difficult
to obtain a correlation between a concentration of a given compound and nitrogen status of a plant. An amide that can be used in this type of analysis is
arginine, which was correlated with nitrogen status of Thompson Seedless
grapes by Kliewer and Cook (1974). Nevertheless, the nitrate-N concentration
in petioles is used for routine field evaluations because of the relatively greater
ease of extraction and analysis of this nitrogen fraction in the laboratory.

Table 3-1.   Nitrogen content of corn plant tissues as influenced by applied nitrogen.

Sampling date	Kilograms of nitrogen applied/hectare				
	0	112	196	280	364
	Whole leaf -- % N[a]				
14 July	2.75	3.37	3.45	3.57	3.72
28 July	1.91	2.55	2.75	2.91	2.98
17 August	2.23	2.55	2.59	2.66	2.65
	Mid-rib -- NO_3-N, ppm[a]				
14 July	1,131	10,140	22,100	12,090	14,495
28 July	351	520	1,345	3,555	4,810
17 August	163	182	240	377	526
	Basal stem -- NO_3-N, ppm[a]				
14 July	2,568	7,865	11,635	13,845	18,915
28 July	878	2,659	6,981	13,143	15,275
17 August	403	2,216	3,562	5,122	5,889

Source: Rauschkolb et al., 1974a.

[a] Values are means of four replications.

Table 3-2.   Nitrogen content of sorghum tissue as influenced by applied nitrogen.

Sampling date	Kilograms of nitrogen applied/hectare					
	0	56	112	178	224	280
	Whole leaf--% N					
6 August	2.00	2.30	2.71	3.04	3.00	3.30
20 August	1.39	1.66	2.21	2.40	2.67	2.87
3 September	1.11	1.61	2.10	2.21	2.54	2.71
17 September	1.22	1.26	1.61	1.76	2.00	2.27
	Mid-rib -- NO_3 -N, ppm[a]					
6 August	364	529	3,215	4,853	9,533	10,660
20 August	148	148	243	494	849	2,427
3 September	117	117	108	135	238	555
17 September	78	135	100	109	82	143
	Basal stem -- NO_3 -N, ppm[a]					
6 August	447	1300	3,449	5,824	9,412	14,300
20 August	225	208	728	1,993	4,117	7,887
3 September	144	139	182	325	2,149	3,467
17 September	148	144	200	295	1,049	2,253

Source: Rauschkolb et al., 1974b.

[a] Values are means of three replications.

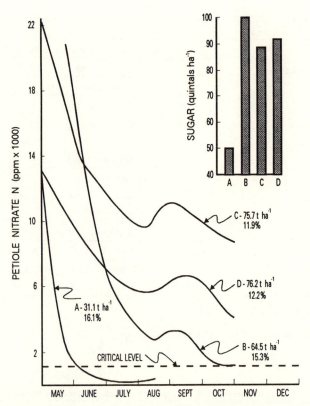

Figure 3-3 Nitrate-N concentration in dried sugarbeet petioles as a function of time and yield of sugar for four different fields (A, B, C, and D) with different nitrogen fertility levels. (Ulrich and Hills, 1973. Copyright © 1973 by the Soil Science Society of America. Reprinted by permission.)

Nitrogen Assimilation

In early investigations of nitrogen fertilizer sources for plants, it was assumed that plants would demonstrate preferences for one form or another of nitrogen supplied to them. Over the years, investigators (Virtanen and Linkola, 1946; Ghosh and Burris, 1950; Hattori, 1957; Kirby and Mengel, 1967; Rauschkolb, 1968) have found that while many forms of nitrogen may be absorbed by plants in varying degrees, the principal forms used by plants are ammonium and nitrate. Because of rapid transformation that occurs in soils, nitrate becomes the predominate form of available soil nitrogen. Nevertheless, most nitrogenous metabolic intermediates can be found in low concentration in the soil. These compounds are of little consequence in relation to nitrogen supply to plants or the movement of nitrogen through soils except where very high loading rates of organic nitrogen have been added.

A schematic showing the principal processes involved in the assimilation of nitrogen is shown in Fig. 3-7. An important first step in the process is the reduction of nitrate through a series of steps to ammonium ion. This is carried

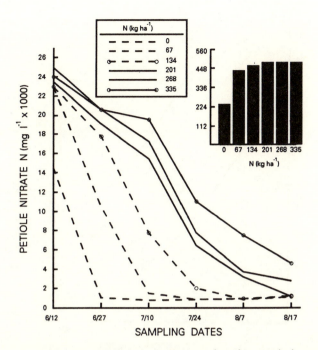

Figure 3-4 Nitrate-N concentration in potato petioles (dry weight basis) as a function of amount of applied nitrogen fertilizer and time and yield of potatoes. (Jones and Painter, 1974.)

out by a series of reductase enzymes with each being specific for a given intermediate step. Once in the ammonium form, nitrogen then enters into a series of metabolic reactions that prevents the accumulation of ammonium and allows the metabolic synthesis of amino acids and proteins. Figure 3-7 also shows the relationship of amino acid from one of the Krebs cycle intermediates. This is one of the major pathways for respiration in plants.

Although Veits and Hagerman (1971) have pointed out that nitrate reductases and other reductase enzymes have been found in various tissues of different plants, the bulk of the reduction takes place in the leaf tissue adjacent to the veins in the leaf blade. Characteristically, the stem, petiole, or midrib, and veinal tissue of a leaf may contain several thousand ppm of nitrate-N, whereas the interveinal tissue of a leaf will contain a few hundred ppm. Exceptions to this occur where the cofactors for the reductase enzymes are limiting or where environmental factors reduce the activity of the reductase enzymes. This sharp and characteristic demarcation of nitrate concentration between leaf and conductive tissue has even been used as a diagnostic technique for molybdenum, which is the cofactor for nitrate reductase.

In several fields in California where molybdenum deficiency had been suspected, the author (RSR) (unpublished data) has found nitrate accumulations in leaf tissue similar to that found in the conductive tissue. In areas of the field where there were no apparent signs of molybdenum deficiency, the normal differences in nitrate concentration between the two tissues were observed. Because

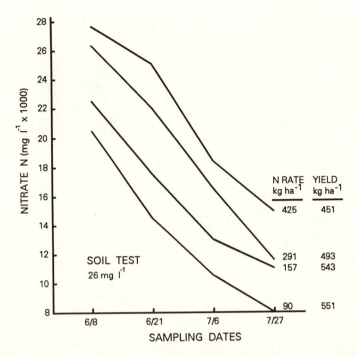

NITRATE N (mg l⁻¹ × 1000)

SAMPLING DATES

N RATE kg ha⁻¹	YIELD kg ha⁻¹
425	451
291	493
157	543
90	551

SOIL TEST
26 mg l⁻¹

Figure 3-5 Influence of initial soil test nitrogen level and amount of applied nitrogen fertilizer on nitrate-N concentration in potato petioles (dry weight basis) and yield of potatoes. (Jones and Painter, 1974.)

of the essentiality of molybdenum for nitrate reductase activity, when a deficiency occurs the enzyme is unable to carry out the reduction rapidly enough to prevent nitrate accumulations within the leafy tissue. As a result, this technique can be used as an indirect indicator of molybdenum deficiency.

The capability of the plant to utilize stored nitrogen in proteins for the formation of amino acids is an important reaction in terms of the overall nitrogen economy of the plant. In senescence of the plant, the stored protein is degraded, and may be used in the development of reproductive tissue, or reconverted into protein in the scaffold of perennial plants where it is stored until it is reactivated for the next season's growth. As much as 80 percent of the leaf proteinaceous nitrogen has been shown to be translocated out of the leaf back into the scaffold of deciduous trees. In the spring the nitrogen becomes available for the early leaf and bud development.

Because of the reduced nature of nitrogen in proteins it might be assumed that ammonium would be the preferred form for plant uptake and assimilation. Ammonium requires less expenditure of energy by the plant than does nitrate, which must be reduced before it can be utilized in metabolic pathways. It has been shown by Maynard and Barker (1969), and Cox and Reisenauer (1973) that small amounts of ammonium in the culture solution in excess of that utilized in growth actually produce toxic reactions and reduce growth areas. Other investigators, including Vines and Wedding (1960) and Kirby and Mengel (1967),

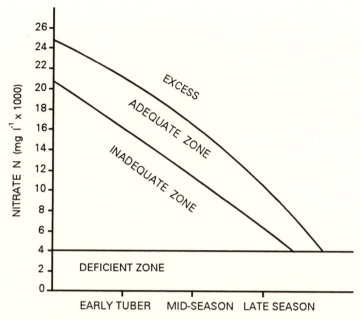

Figure 3-6 Suggested nitrate-N levels in potato petioles through the growing season (Jones and Painter, 1974.)

found that ammonium could effectively inhibit respiration by inhibiting the energy transfer enzymes that are essential for many metabolic processes. However, plants do contain low levels of ammonium in their tissue, and some investigators consider the presence of ammonium as essential to plants.

A steady-state nitrogen concentration solution culture was used by Cox and Reisenauer (1973) to examine yield and nitrogen uptake by wheat. At approximately 0.6 ppm ammonium-N in solution the yield of wheat reached a maximum. With increasing concentrations to approximately 1.2 ppm yield of wheat then decreased. Increasing nitrate-N in the culture solution to about 1.4 ppm increased wheat yield. There was no decline in yield even though the amount of nitrogen supplied as nitrate-N was greater than when ammonium-N was the sole nitrogen source. When approximately 0.6 ppm ammonium-N was added to a cultural solution containing 2.8 ppm nitrate-N, yield was increased nearly 50 percent over that obtained by the nitrate source alone. As the ammonium concentration increased to 1.2 ppm, the yield decreased to the yield level achieved when nitrate was the sole source of nitrogen. These data support conclusions that ammonium toxicity can occur in plants but does not explain the mechanism by which it occurs.

Banding of ammonia close to plant roots or placement of ammonium fertilizers in broad bands below seedlings or germinating plants have also been observed to result in ammonium toxicities of plants. It sometimes may be difficult in the field to separate the effect of high pH from specific ammonium toxicity as a result of the banded anhydrous ammonia. In the field situation one

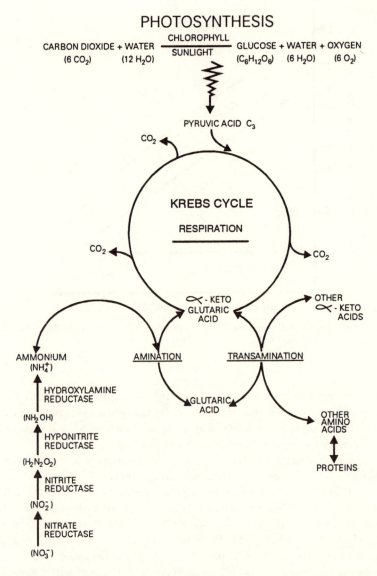

Figure 3-7 Schematic showing the major metabolic processes for nitrogen assimilation in plants. (Conn and Stumpf, 1963. Copyright © 1963 by John Wiley & Sons, Inc. Reprinted by permission.)

needs to be especially aware of the possibility of ammonium toxicity and place the ammonium fertilizer in such a manner as to prevent this effect.

Deficiency Symptoms

The ability of the plant to mobilize and translocate stored nitrogen provides clues as to the visible symptoms the plant expresses when suffering a deficiency. There are three levels at which nitrogen deficiency symptoms that may be observed in the field will lead to the diagnosis of a deficiency. An individual leaf will show a general yellowing. The leaf becomes more yellow (chlorotic) as the severity of nitrogen deficiency increases. The yellowing is the result of inadequate nitrogen for the formation of chlorophyll pigment that gives the plant its green color. As the deficiency becomes more severe, protein is degraded and translocated causing the leaf tissue to become necrotic (the tissue dies). Some plants may even have a characteristic pattern for development of the chlorotic areas. For example, corn exhibits a "V" shaped necrotic area starting from the tip of the leaf with the top of the "V" pointed toward the plant.

The second level of nitrogen deficiency is overall plant appearance. The plant is stunted and spindly since rapid growth and development is inhibited by the lack of adequate nitrogen. Since nitrogen is mobile within the plant, the lower leaves are the first to turn yellow and become necrotic. If the deficiency is not too severe, the upper leaves may be a light green with the lower leaves chlorotic. As the deficiency becomes more severe, chlorosis appears higher up on the plant and eventually the plant becomes entirely chlorotic and dies.

The third level of deficiency is related to the uniformity of soil and water application in a management unit. If the soils are uniform, a general light green appearance will be observed for all the plants as the plants become slightly nitrogen deficient. The inexperienced observer may not detect such a mild deficiency. In a management unit that does not have uniform soil characteristics or water movement characteristics, chlorotic and stunted plants will appear in parts of the field. These areas will be closely associated with textural difference within the management unit. At low fertility levels, the entire management unit may be uniformly chlorotic even though it contains nonuniform soils. In orchards or vineyards the observable nitrogen deficiency symptoms are more subtle, requiring much greater experience in their detection. Stored nitrogen plays such a large roll in the growth and development of new tissue that there are often delayed effects of poor nitrogen nutrition for a given year. However, sustained nitrogen deficiency will result in much the same observable deficiency symptoms as in annual plants. An experienced observer can detect a much slower rate of shoot and leaf development and perhaps some mild chlorosis.

ROOTING CHARACTERISTICS

The function of a root system is to absorb nutrients and water from the soil matrix, which are then transported to the remainder of the plant. Except for major differences between monocotyledonous and dicotyledonous plants, the internal root anatomy does not vary greatly from one plant species to the next.

The mineral elements absorbed by plants are generally present in extremely low concentrations as inorganic ions in the soil solution. As the soil solution near the root becomes depleted of nutrients, diffusion and mass flow replenish their supply. Root systems have an astonishingly large surface area. This is brought about by means of progressive branching and root extension, which facilitates absorption of mineral elements.

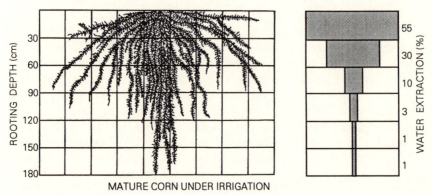

Figure 3-8 Root distribution and water extraction patterns for mature irrigated corn (*Zea mays*).

Types of Root Systems

Great variations in branching, lateral development, and depth of development for root systems occur for different plant species. Examples of root systems for various crops are shown in Figs 3-8 through 3-15. These show the depth of development for representative agronomic crops and deciduous trees with root systems ranging from 4 to 12 ft (1.2 to 3.6 m) deep. It is well known that many vegetable crops have much shallower root systems and more limited lateral development. Rooting depths for many vegetable crops do not exceed 60 to 90 cm and later development may have a radius of less than 30 cm.

It is common practice for growers to use deep-rooted plants as a crop following shallow-rooted plants to help extract residual soil nitrogen from deeper in the soil profile in order to minimize the fertilizer needed to grow the crop. This is especially true where the crop rotation involves vegetable production. Because

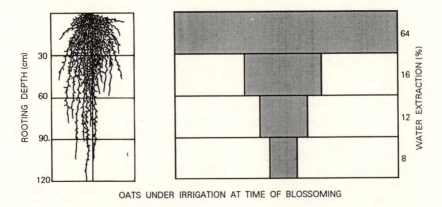

Figure 3-9 Root distribution and water extraction patterns for irrigated oats (*Avena sativa*) at time of blossoming.

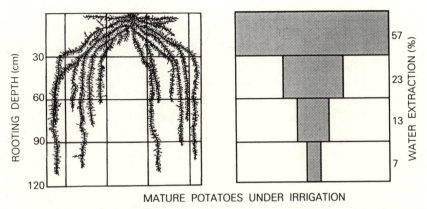

Figure 3-10   Root distribution and water extraction patterns for irrigated mature potatoes (*Solanum tuberosum*).

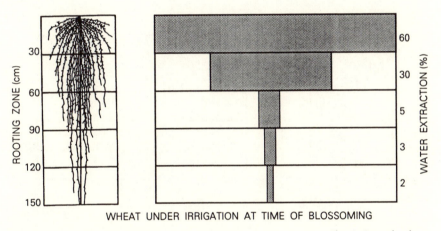

Figure 3-11   Root distribution and water extraction patterns for irrigated wheat (*Triticum aestivum*) at time of blossoming.

of the shallow-rooted nature of vegetables and the high rates of nitrogen applied there is usually sufficient carryover to reduce the amount of fertilizer that needs to be applied to the following crop.

It should be evident that the extent of root systems of crops can have a major impact on determining where and when nitrogen fertilizer should be applied. Furthermore, the type of root system and rate of development during the growing season will affect the uptake of nitrogen. There have been numerous studies that show that the availability of nitrogen to the plant is dependent upon the placement of nitrogen in a position where it is accessible by plant roots. This may either be the result of mechanical placement or may be accomplished by movement of nitrate-N with irrigation to plant roots.

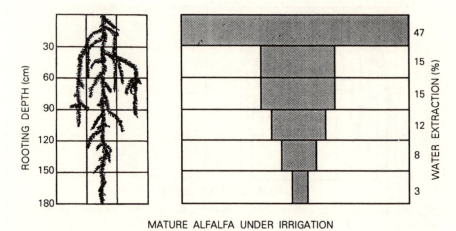

Figure 3-12 Root distribution and water extraction patterns for irrigated mature alfalfa (*Medicago sativa*).

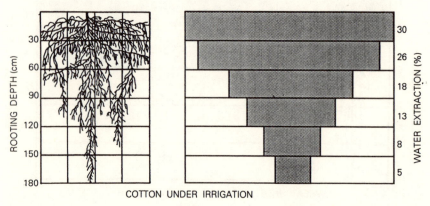

Figure 3-13 Root distribution and water extraction patterns for irrigated cotton (*Gossypium*).

Nitrogen and Water Use

As indicated above, the depth, lateral extent, and density of the root system play a significant role in determining the potential nitrogen carryover, and subsequent nitrogen pollution potential. Scherty and Miller (1972) have shown that large amounts of residual nitrate-N may be removed from soils by alfalfa, which is a deeply rooted plant. This study points out the value of using deep-rooted plants for extracting nitrogen from soils after growing shallow-rooted crops.

Gass et al. (1971) have shown that the rate of root development and the soil fertility level influence the amount of nitrogen absorbed from different soil depths. In studies using isotopic nitrogen placed at various depths below the soil surface, they found that 92 percent of the nitrogen was taken up from the

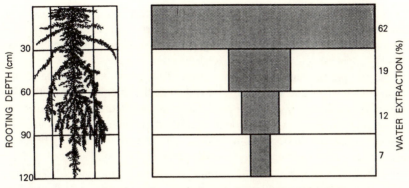

THREE - MONTH - OLD SUGAR BEETS UNDER IRRIGATION

Figure 3-14 Root distribution and water extraction patterns for irrigated three-month-old sugar beets (*Beta vulgaris*).

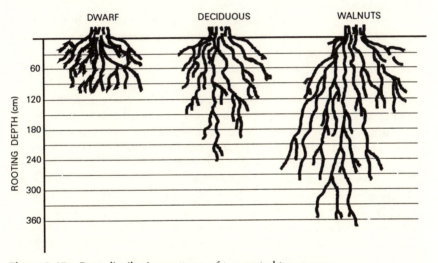

Figure 3-15 Root distribution patterns of tap-rooted tree crops.

zero to 60-cm depth at the highest residual nitrogen level (582 kg/ha). In plots that contained lower residual nitrogen levels (409 and 468 kg/ha), 30 and 66 percent, respectively, of the isotopic nitrogen was absorbed from below the 120-cm depth. These studies serve to point out the dynamic nature of root development and the impact of that development on nutrient uptake.

Similar studies have been conducted on water extraction from root zones (see Figs 3-8 through 3-15). The interaction between water extraction from soils and absorption of plant nutrients makes it necessary to consider the influence of moisture distribution in the soil profiles on nutrient uptake. In a normal irrigation cycle the soil profile to the bottom of the root zone is filled with water to the extent of the water-holding capacity of the soil. A rule of thumb for extraction of soil water by plants was proposed by Israelsen and Hansen (1967). Approximately 40 percent of the water will be extracted from the top one-quarter of the root zone. Approximately 30, 20, and 10 percent are extracted from the remaining second, third, and bottom quarters of the root zone, respectively. This corresponds relatively well with the nitrogen extraction pattern observed by Gass et al. (1971) for corn.

As the soil profile begins to dry as a result of water extraction by the plants, the amount of water and nutrients removed from the drier portion of the soil profile is reduced. It is a common observation in irrigated areas that soils dry to a depth of 15 to 30 cm with prolonged periods between irrigations. Relatively little nutrient uptake can occur from the drier zones where root activity becomes limited. Such surface drying suggests that fertilizer should be placed at least 15 cm below the soil surface. If the nitrogen is uniformly distributed throughout the profile because of the method of fertilizer application, incorporation of crop residues, or as a result of mineralization of soil organic matter, which character-istically occurs in the upper 30 cm of the soil, then the greatest amount of available nitrogen is generally in the top 30 to 60 cm of soil unless irrigation has redistributed the nitrate-N deeper in the soil profile.

In a soil having a relatively low water-holding capacity and high infiltration rate, fertilizer applied just below the soil surface at planting time may remain just out of reach of the plant root system of an annual crop through a combi-nation of slight overirrigation and the rate of root development during the sea-son. Each irrigation can move the nitrate slightly deeper in the profile, thus preventing the root system from reaching the major nitrogen supply. Even though an adequate amount of nitrogen fertilizer had been applied at the begin-ning of the season, the plant may never have had an opportunity to utilize much of the nitrogen. As far as the plant was concerned, there existed inadequate nitrogen for its growth and development.

Plants with shallower rooting characteristics are more likely to be affected by leaching of the nitrogen below the point where it is accessible. For deeply rooted plants, the rate of root development may be more rapid than the rate of downward movement of nitrate by irrigation water. In fine textured soil, the rate of nitrate movement through the soil profile under normal irrigation regimes is not so high that the plant is unable to catch up to the nitrogen supply.

DEVELOPMENT OF TOPS

The growth and development of plant tops have been characterized to a greater extent than for root systems. Figure 3-16 shows the general manner in which a plant develops. As the plant develops from a seedling into a mature plant, the plant enters what is referred to as the grand period of growth. As the plant approaches maturity, the rate of growth slows and the reproductive portion of the plant develops. In some crops part of the plant may be harvested during the rapidly developing vegetative stage instead of waiting for the plant to reach the reproductive stage. Even though the vegetative portion of the plant may be harvested, the cultural practices used to grow the crop may still dictate that growth should be slowed prior to harvest. Sugarbeets and several vegetable crops are cases in point.

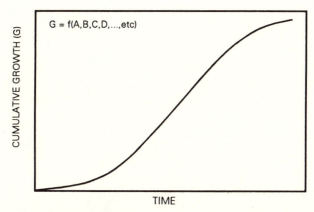

Figure 3-16 The relative cumulative growth of a plant with time is influenced by several factors each of which may influence growth by changing over a continuous range of possible levels and their interaction.

A continuum of possible yield levels exists, each of which is subject to the effects of numerous factors such as climate, water supply, insects, diseases, and other variables. The yield potential of a crop has been defined as the yield expected assuming an adequate fertility level. The nitrogen requirement for plants grown under high yield potential conditions is greater than the nitrogen required for plants with low yield potential. This necessarily assumes there is a direct relationship between the growth of the plant and the amount of nitrogen taken up by the plant during the growing period.

NITROGEN UPTAKE, RESPONSE AND REMOVAL

Nitrogen Uptake

The relationship between dry matter accumulation and cumulative nitrogen uptake has been determined for several crops. Figures 3-17 through 3-20 show examples of this relationship for cantaloupe, lettuce, rice, and sorghum. Similar types of information are also available for many other crops. The four crops referred to have diverse growth characteristics and the correlation observed between dry matter production and nitrogen uptake supports the principle of a direct relationship between plant growth and nitrogen uptake. These relationships provide useful information regarding the time during the plant growth period when nitrogen is required in greatest quantities. Such information assists in timing the application of fertilizer so that adequate nitrogen is present prior to the time when the largest demand occurs. Only about 15 percent of dry matter production and total nitrogen uptake occurred during the first 90 days of lettuce plant growth (Fig. 3-18). During the next 30-day period, the remaining 85 percent of growth and nitrogen uptake occurred. In contrast, about 60 percent of the dry matter accumulation for rice (Fig. 3-19) occurred within the first 90 days

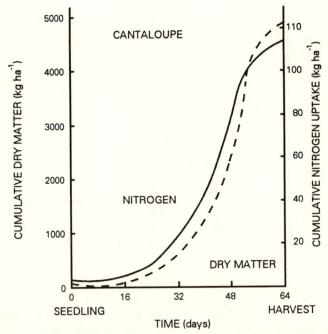

Figure 3-17 Cumulative dry matter and nitrogen uptake for fruit and vines of cantaloupe (*Cucumis melo cantalupensis*) from seedling to harvest. (B.R. Gardner, University of Arizona, Yuma Valley Experimental Farm, personal communication, 1977.)

after emergence with approximately 40 percent of the total growth occurring in the next 70 days. Nitrogen uptake again closely follows the biomass accumulation. These and similar data indicate how critical it is to provide nitrogen for the rapid growth period of plants when it is needed and how essential it is to know the growth characteristics of the plant in order to adjust the nitrogen supply to meet the plant demand. These curves were developed under conditions where nitrogen was neither limiting nor excessive.

Nitrogen Response

Another characteristic of plant growth in relation to nutrient supply can be observed in Figure 3-21. This figure shows the relationship between cumulative growth and different levels of nutrient supply at selected stages of growth. The shaded portion of the curve depicts the typical nutrient response curve.

A nutrient response curve for sorghum is shown in Fig. 3-22. In this case the soil was highly nitrogen deficient but was still capable of supplying nitrogen for a yield of about 2400 kg/ha. The points in the graph are the means of four replications and the curve is the best statistically fitted curve of all the data points. A maximum of about 7000 kg/ha was reached with approximately 200 kg/ha of applied nitrogen. The principal features of this curve to note are that some yield was obtained even when no nitrogen was applied and that

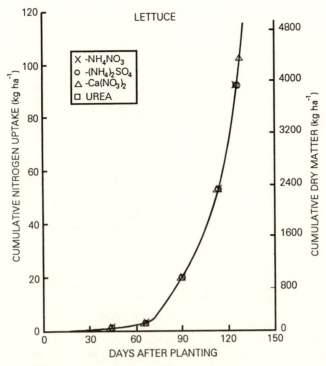

Figure 3-18 Cumulative nitrogen uptake and dry matter for lettuce (*Lactuca*) with four nitrogen sources from planting in mid-November to harvest in March. (B.R. Garner, University of Arizona, Yuma Valley Experimental Farm, personal communication, 1977.)

incremental additions of nitrogen increased yield until a maximum level was achieved.

In situations where excessive or adequate nitrogen levels may be present in soils as carryover from a previous crop or from organic matter decomposition, the yield achieved even with zero applied nitrogen may approach maximum. Figure 3-23 shows the yield of barley with increasing nitrogen supply. In this case, yield decreased with each increment of nitrogen fertilizer applied. In this study the soil nitrate-N concentration in the top 60 cm averaged 131 ppm, which is comparable to an available nitrogen supply of approximately 600 kg/ha. That amount of nitrogen far exceeds that required to produce maximum yield. These data serve to point out that excessive amounts of nitrogen, either residual in the soil or as a result of fertilizer application, can reduce yields. Similar results have been observed on other crops as shown previously in this chapter (Figs 3-4 and 3-5). The adverse impact of excessive soil nitrogen on yield of many crops indicates the importance of accounting for available soil nitrogen when planning a nitrogen fertilization program.

Not all plant species respond in the same way to excessive amounts of nitrogen. Some have a rather broad plateau between the point on the response curve where maximum yield is achieved and where yield reductions occur from

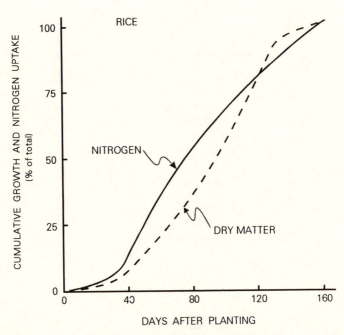

Figure 3-19 Cumulative dry matter production and nitrogen uptake as a percent of the total for rice (*Oryza sativa*). (Abridged from Mikkelsen and Patrick, 1968. Copyright © 1968 by the Soil Science Society of America. Reprinted by permission.)

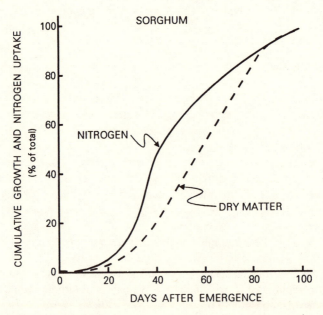

Figure 3-20 Cumulative dry matter production and nitrogen uptake as a percent of the total for sorghum. (Abridged from Roy and Wright, 1973 and 1974. Copyright © 1973 and 1974 by the American Society of Agronomy. Reprinted by permission.)

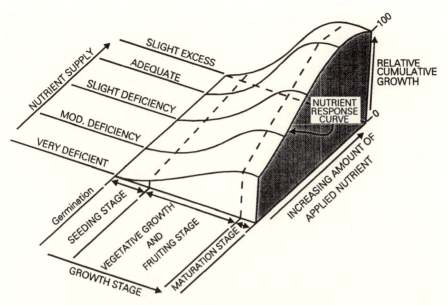

Figure 3-21 Diagram of the relationships between relative growth at various nutrient levels and the yield response curve resulting from applied nutrient.

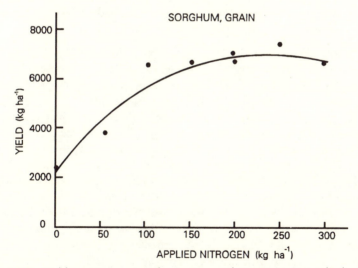

Figure 3-22 Yield response curve for incremental increases in applied nitrogen fertilizer. (Rauschkolb et al., 1974b.)

excessive nitrogen. It has been observed that field crops such as corn, sorghum, pasture or forage grasses, as well as certain vegetables such as lettuce, spinach, celery, and onions may withstand rather large excesses of applied nitrogen without observable yield reductions. Other crops such as sugarbeets, cotton, grapes, and apricots are fairly sensitive to excessive amounts of nitrogen.

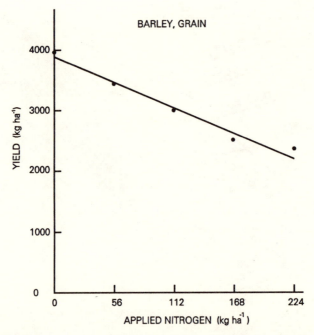

Figure 3-23 Influence of applied nitrogen on yield of barley (*Hordeum vulgare*) grown on a soil with a relatively high level of nitrate-N in the soil. (Rauschkolb, Kite, and Sharrat, University of California–Davis, unpublished data, 1973.)

The information in Fig. 3-24 presents data from an extensive field study funded by the National Science Foundation, Research Applied to Nation Needs program (Pratt, 1979a) that examined the relationship between yield, nitrogen supply, and residual soil nitrogen using corn as the test crop. The study was conducted using isotopically labeled nitrogen to distinguish between soil and fertilizer nitrogen. The results depicted in Fig. 3-24 are for the first year of the study conducted at one of two locations, but are representative of the results obtained for both locations after 5 years. The field was extremely nitrogen-deficient as indicated by the low yield when no nitrogen was added. Substantial yield increases occurred with each increment of added nitrogen until maximum yield was obtained. Nitrogen additions beyond that required for maximum yield resulted in a slight decline in yield. Even though maximum grain yield was achieved with 224 kg/ha of N, the above-ground portion of the plants continued to accumulate nitrogen from the next increment of fertilizer as well. Additional increments of nitrogen resulted in no detectable increase in the total amount of nitrogen taken up. The maximum amount of soil nitrogen taken up by the plants coincides with the point at which maximum yield was achieved, i.e. 224 kg/ha of N. As additional fertilizer nitrogen was applied beyond the point where maximum yield was achieved, decreasing amounts of soil nitrogen were taken up by the plant. Another of the curves shows that the maximum amount of available soil and fertilizer nitrogen found in the grain yield was obtained at maximum yield. The curve showing available soil and fertilizer nitrogen is the

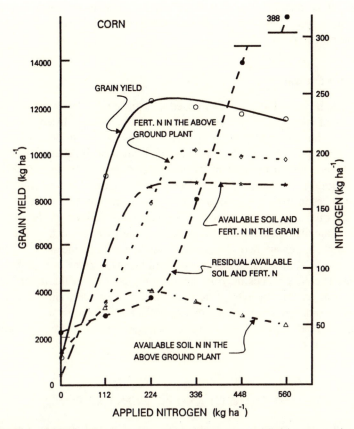

Figure 3-24 Relationships between yield response to added nitrogen, the uptake of soil and fertilizer nitrogen, and residual nitrogen after harvest of the corn (*Zea mays*) grain. (P.R. Stout, University of California–Davis, unpublished data, 1974.)

total available nitrogen found in the soil to a depth of 3 m after harvest of the crop.

There are important interactions to be considered and interpretations to be made in relation to the residual soil nitrogen, fertilizer practices, and nitrogen pollution. Even with no applied nitrogen, the residual soil nitrogen level was nearly 50 kg/ha. There was an increase of only 20 to 25 kg/ha (the equivalent of about 10 percent of the applied nitrogen) in the amount of the residual nitrogen at harvest at the point where maximum yield was obtained. The plant was not only utilizing applied fertilizer nitrogen at a relatively high efficiency to this point, but it was also extracting a greater amount of soil nitrogen. This relationship between maximum yield and minimal nitrogen carryover, as shown in the slight increase in residual available nitrogen, has been demonstrated in each of the 5 years of the field study at both locations.

Studies conducted without the benefit of isotopic tracers have shown similar relationships, indicating that adequate but not excessive amounts of fertilizer for production of near-maximum yields result in minimal carryover of residual nitrogen. It is equally evident that increases in applied nitrogen beyond the point where maximum yield is obtained contribute significantly to the nitrate pollution potential.

Nitrogen Removal

Nitrogen removal values have been used as a means of evaluating the relative efficiency of nitrogen uptake for crops. In Table 3-3 are listed the nitrogen contents of the harvested portion of several crops. Where information is available on other portions of the crop, these data are also included. The nitrogen content values are most important for making comparisons between yields, since yield potential may change from one location to another. These values provide a means of estimating the total amount of nitrogen removed and the apparent efficiency of fertilizer use. (Apparent efficiency is defined as the amount of nitrogen contained in the harvested portion of the plant divided by the amount of fertilizer nitrogen applied.) Such values can be used as a means for comparing efficiencies of crop production practices from one year to the next. Even though apparent efficiency values may not be accurate in terms of the true efficiency, one can still attempt to manage fertilizer, water resources, and other cultural practices in order to achieve high apparent efficiency for a given location. True efficiency would require accounting for nitrogen removed from soil reserves and residual sources.

MODELING PLANT GROWTH AND NUTRIENT RESPONSE

Historically, the approaches to evaluation of plant growth responses to nutrient supply have had both simplicity and utility. In spite of the fact that numerous other factors are also known to influence plant growth, a distinguishing feature of these techniques has been the uniformity of growth coefficients that were determined over a wide range of conditions.

In the early 1900s a German scientist, E.A. Mitscherlich, proposed a mathematical equation that related growth of plants to nutrient supply. A similar equation was developed independently by an American, W.J. Spillman, a few years later. Both equations can be expressed in the same form as

$$\log (A - y) = \log A - 0.301x$$

where A is the maximum possible yield, y is the observed yield for a given nutrient level, 0.301 is a constant depending upon the nature of the growth factors, and x is the number of units of the growth factor required to give an increase in yield. Over the years this concept has been used to evaluate plant responses to nutrient supply. It has also been modified to express growth of the plant as a function of less than optimum amounts of the growth factor. This approach, while simplistic, has proved to be useful in evaluating plant nutrient responses.

Table 3-3.   Nitrogen content and removal for selected crops grown at the indicated yield[a].

Selected crops	Nitrogen content	Crop yield[b,c]	Nitrogen removal[c]
	----% kg/ha----	----% kg/ha----	—%kg/ha—
Field Crops:			
Barley, grain	1.89	4,500	85
straw	0.91[d]	4,400[d]	40[d]
Beans, dry, beans	3.46	3,000	104
stalks	1.00[d]	6,000[d]	60[d]
Corn, grain	1.47	10,000	147
stover	0.30	25,000	75
Cotton, lint	---[e]	1,000	---
seed	3.50	1,900	67
burrs, leaves & stems	1.50[d]	3,000[d]	45[d]
Oats, grain	1.86	3,000	56
straw	0.70[d]	4,000[d]	28[d]
Rice, grain	1.40	6,000	84
straw	0.17	7,500	13
Safflower, grain	3.06	3,000	92
stalks	0.70[d]	4,000[d]	28[d]
Sorghum, grain	1.86	4,500	84
stover	1.10[d]	4,000[d]	38
Sugarbeets, roots	0.20	56,000	112
tops	0.47	30,000	141
Wheat, grain	1.76	4,500	79
straw	0.93[d]	35,000[d]	33[d]
Hay and Silage:			
Alfalfa, hay	2.90	13,000	378
Mixed grass, hay	2.08[d]	4,500[d]	94[d]
Silage, corn & sorghum	0.39	56,000	218
Irrigated pasture, mixed species	1.50[d]	4,500[d]	68[d]
Range, mixed species	1.50[d]	600[d]	9[d]
Fruits and Nuts:			
Apple, fruit	0.80	36,000	29
leaves & wood	---	---	65[f]
Apricot, fruit	0.19	18,000	34
leaves & wood	---	---	80[f]
Cherry, fruit	0.21	9,000	19
leaves & wood	---	---	---
Grapes, berries	0.10	22,400	22
leaves & vines	---	---	60[f]
Peach, fruit	0.14	36,000	50
leaves & wood	---	---	60[f]
Pear, fruit	0.11	34,000	25
leaves & wood	---	---	50[f]
Plum, fruit	0.24	18,000	54
leaves & wood	---	---	35[f]
Prune, fruit	0.24	18,000	54
leaves & wood	---	---	---
Almonds, meats	2.98	2,000	60
hulls	---	---	---
Pecans, meats	2.40	2,800	67
hulls	---	---	---
Citrus and Subtropical Fruits:			
Grapefruit	0.16	24,000	38
Lemon	0.19	30,000	57
Orange, fruit	0.21	18,000	38
leaves & wood	---	---	35[f]

Continued--.

Table 3-3 (continued)

Selected crops	Nitrogen content	Crop yield[b,c]	Nitrogen removal[c]
	----% kg/ha----	----% kg/ha----	—% kg/ha—
Tangerine	0.13	12,000	16
Avocado	0.35	5,800	20
Fig	0.19	6,500	12
Olive	0.19	4,600	9
Vegetables:			
Artichokes	0.43	7,500	32
Asparagus	0.54	3,800	17
Beans, snap	0.43	12,000	52
Beets, table, roots	0.30	35,000	105
tops	0.65	23,000	152
Broccoli	0.58	10,000	58
Brussels sprouts	0.78	15,000	117
Cabbage	0.22	30,000	67
Cantaloupes, melons	0.16	22,400	36
vines	---	---	60[f]
Carrots, roots	0.18	42,000	76
tops	0.50	20,000	98
Cauliflower	0.35	15,000	56
Celery	0.18	75,000	135
Corn, sweet, ears	0.64	12,000	77
stover	0.40	15,000	60
Cucumbers, cukes	0.10	30,000	30
vines	---	---	60[f]
Garlic	1.02	15,000	153
Lettuce	0.16	30,000	48
Onions	0.24	36,000	86
Peas, peas	1.01	3,600	36
vines	0.25	18,000	45
Pepper, bell	0.19	24,000	46
Potatoes, tubers	0.34	45,000	153
vines	---	---	100[f]
Radishes, roots	0.18	22,400	40
tops	0.45	15,000	68
Spinach	0.51	22,400	114
Strawberry, fruit	0.15	44,000	66
vines	2.13[d]	4,000	85[d]
Vegetables:			
Tomatoes, fruit	0.18	56,000	101
Turf-grasses:			
Bent grass	3.08[d]	5,000[d]	154[d]
Bermuda grass	2.81[d]	9,000[d]	253[d]
Kentucky blue grass	2.77[d]	5,000[d]	139[d]

[a] Data taken from: Beeson, 1941; McVickar et al., 1963; National Plant Food Institute, 1966; Beutel and Uriu, University of California, personal communication; Watt and Merrill, 1963; and the Western Fertilizer Handbook, 1980.

[b] Yields will vary with location, year and management. Yields are based on normal field harvested weights unless otherwise stated.

[c] Values are rounded off for convenience of reporting.

[d] These values are based on dry weights.

[e] Dashed line indicates no reported values.

[f] Values for removal are adapted from the Western Fertilizer Handbook (1980), no data was provided to calculate N content of the tissue.

Table 3-4. Some process-level crop simulation efforts.

Research group	Institutions	Model name	Species	Processes treated
Acock, B., V.R. Reddy, F.D. Whisler, D.N. Baker, J.M. McKinion, H.F. Hodges, and K.J. Boote	USDA-ARS, Mississippi State U., and U. of Florida	GLYCIM	Soybean	Photosynthesis, respiration, transpiration, growth, and morphogenesis. Incorporates RHIZOS
Allen, J., and J.H. Stamper	U. of Florida	CITRUSIM	Citrus	Photosynthesis
Angus, J.F., and H.G. Zandstra	CSIRO (Australia) and International Rice Research Institute	IRRIMOD	Rice	Growth, phasic development, soil water flow, soil nitrogen, transpiration, and evaporation
Arkin, G.F., J.T. Ritchie, and R.L. Vanderlip	Texas A&M U., USDA/SEA, and Kansas State U.	SORG	Sorghum bicolar	Photosynthesis, respiration, transpiration, and evaporation
Baker, D.N., J.R. Lambert, and J.M. McKinion	USDA/SEA (Mississippi) and Clemson U.	GOSSYM	Cotton	Photosynthesis, respiration, growth, and morphogenesis. Incorporates RHIZOS
Baker, D.N., D.E. Smika, A.L. Black, W.O. Willis, and A. Bauer	USDA/SEA (Mississippi, Colorado, and North Dakota)	WINTER WHEAT	Wheat	Photosynthesis, respiration, transpiration, growth, and morphogenesis. Incorporates RHIZOS
Brown, L.G., J.D. Hesketh, J.W. Jones, and F.D. Whisler	Mississippi State U.	COTCROP	Cotton	Photosynthesis, respiration, transpiration, runoff, drainage, nitrogen uptake, denitrification, leaching, organogenesis, partitioning, and growth
Childs, S.W., J.R. Gilley, and W.E. Splinter	U. of Nebraska	unnamed	Corn	Photosynthesis, respiration, transpiration, growth, soil evaporation, and soil water flows
Curry, R.B., G.E. Meyers, J.G. Streeter, and H.L. Mederski	Ohio Agriculture Research and Development Center	SOYMOD OARDC	Soybean	Photosynthesis, respiration, translocation, and evaporation

Continued—

Table 3-4 (Continued)

Research group	Institutions	Model name	Species	Processes treated
Duncan, W.G.	U. of Florida and U. of Kentucky	SIMAIZ	Corn	Photosynthesis, processes involved in setting seed number and seed size
Duncan, W.G.	U. of Florida and U. of Kentucky	MIMSOY	Soybean	Photosynthesis, nitrogen fixation, assimilate redistribution, processes for setting seed number and seed size
Duncan, W.G.	U. of Florida and U. of Kentucky	PEANUT	Peanuts	Photosynthesis, nitrogen fixation, processes for setting seed number and seed size
Fick, G.W.	Cornell University	ALSIM	Alfalfa	Photosynthesis defined as crop growth rate, and partitioning
Holt, D.A., G.E. Miles, R.J. Bula, M.M. Schreiber, D.T. Doughtery, and R.M. Peart	Purdue University and USDA/SEA	SIMED	Alfalfa	Photosynthesis, respiration, growth, translocation, and soil moisture uptake
Jones, C.A., and R.T. Ritchie	USDA/SEA (Texas) and IFDC, Alabama	CERES-MAIZE	Corn	Phasic development, morphogenesis, growth, biomass accumulation and partitioning, soil water balance and plant-soil nitrogen status
Kercher, J.R.	Lawrence Livermore Laboratory	GROW1	General	Photosynthesis, transpiration translocation
van Kenulen, H.	Netherlands Agricultural U. (Wageningen)	GRORYZA	Rice	Gross assimilation and respiration

Continued--.

Table 3-4 (Continued)

Research group	Institutions	Model name	Species	Processes treated
van Keulen, H.	Netherlands Agricultural U. (Wageningen) ARIDCROP	ARIDCROP	Natural vegetation in semiarid regions	Photosynthesis, respiration, transpiration, and water uptake
Lambert, J.L., D.N. Baker, and J.M. McKinion	Clemson U. and USDA/SEA (Mississippi)	RHIZOS	Soil	Infiltration, uptake, capillary redistribution, ET, nitrogen transformation, N fertilizer applications
Loomis, R.S., and E. Ng	U. of California-Davis	POTATO	Potato	Photosynthesis, respiration, transpiration, water uptake, growth, development, and senescence
Loomis, R.S., J.L. Wilson, D.W. Rains, and D.W. Grimes	U. of California-Davis	COTGRO	Cotton	Photosynthesis, respiration, transpiration, water uptake, growth, development, flowering, fruit development, senescence, and heat flux
Loomis, R.S., G.W. Fick, W.A. Williams, W.H. Hunt, and E. Ng	U. of California-Davis	SUBGRO	Sugar beet	Photosynthesis, respiration, transpiration, water uptake, growth, plant development, and senescence
Marani, A.	The Hebrew U. of Jerusalem	ELCOMOD	Cotton (Acala)	Photosynthesis, respiration, growth, morphogenesis, ET, nitrogen uptake, and gravitational soil wetting
McMennamy, J.A., and J.C. O'Toole	International Rice Research Institute	RICEMOD	Rice	Photosynthesis, respiration, growth

Table 3-4 (Continued)

Research group	Institutions	Model name	Species	Processes treated
Orwick, P.L., M.M. Schreiber, and D.A. Holt	Purdue University	SETSIM	Setaria	Carbon flow, photosynthesis, respiration, growth, and translocation
Ritchie, J.T., and S. Otter	USDA/SEA (Texas)	CERES-WHEAT	Wheat	Phasic development, morphogenesis, growth biomass accumulation and partitioning, soil water balance, plant nitrogen status
Ryle, G.J.A., N.R. Brockington, C.E. Powell, and B. Cross	Grassland Research Institute (Hurley, Berkshire, England)	Unnamed	Uniculum barley	Photosynthesis, assimilate distribution, and synthetic and maintenance respiration
Weir, A.H., P.L. Bragg, J.R. Porter, and J.H. Rayner	Rothamsted Experimental Station, Letcombe Laboratory, U. of Bristol	ARCWHEAT 1	Wheat	Photosynthesis, phenology, respiration, and dry matter partitioning
de Wit, C.T., R.Broener, and F.W.T. Penning de Vries	Netherlands Agricultural U. (Wageningen)	PHOTON and BACROS	Any crop	Photosynthesis, respiration, transpiration, reserve utilization, water uptake, and stomatal control
Wilkerson, G.G., J.W. Jones, K.J. Boote, K.T. Ingram, and J.W. Mishoe	U. of Florida	SOYGRO	Soybean	Photosynthesis, respiration, growth, senescence, phenology, infiltration, drainage, transpiration

Source: Whistler et al., 1986. Copyright © 1986 by Academic Press. Reprinted by permission.

Plant Growth

There have been three main approaches to plant growth modeling. An approach used by Erickson (1976) simulates growth rate for size, area, length, or volume of an organ or tissue. Such changes are commonly assumed to be nearly linear over a considerable period of time. For example, he reported that the primary root of corn grew at the rate of 2 mm/hr for 3 days. In order to adapt this assumption to the generally sigmoid pattern of plant development, he incorporated a log transformation of dry matter accumulation as a function of time. This provides a straight line that can be used for these types of plant growth models. Another type of model simulates plant responses to environmental factors such as temperature, nutrient, and water supply. Such holistic models look at plant response to variations in the external factors affecting the growth and predict growth with changes in those factors as a function of time. A third approach has been to examine the physiological basis for growth and development and to simulate the major processes as they may be influenced by environmental factors, partitioning of constituents, and their translocation in the plant.

These models provide the basis for simulating the morphological development of the plant. This latter approach, although more difficult to apply, should ultimately permit broader application to a variety of plant species and environmental conditions. As pointed out by Loomis (1976), the integrative physiological models appear to offer a means of dealing with growth quantitatively even within the framework of present limitations in knowledge. Furthermore, they state that empirical plant modeling without careful attention to biological aspects affecting growth may reduce the model to an expensive curve-fitting routine.

Many of the simulation models presented in Table 3-4 have used the integrated physiology approach in which photosynthesis, respiration, nutrient uptake, and water supply are simulated with respect to concentration, flux, and climatic effects. Some of the models are in the developmental stage and consequently are not available for general use. Most assume an adequate supply of water and nutrients, although attempts are being made to deal with soil–plant–water interrelationships by incorporating soil-nitrogen transformations and plant–water relations into a more comprehensive simulation model.

4

Irrigation Systems

In any type of agricultural production the fate of nitrogen is inescapably linked with that of water. In irrigated agriculture the fate of water is controlled in turn by the method of irrigation employed. Inherent differences in the way water is applied for each method of irrigation influence solute movement and replenishment of soil-water supply. Uniformity of water distribution, direction of water movement, efficiency of water application, and irrigation efficiency determine the frequency of water application and the amount of water to be applied in any given irrigation. Because of these differing characteristics, the method of irrigation has a major impact on nitrogen availability to the plant through its influence on direction of movement, amount of drainage below the root zone, redistribution within the root zone, and denitrification losses. Furthermore, the amount, frequency, and quality of water moving through the soil influence its chemical and physical characteristics, which may then be reflected in plant growth and the crop's use of water and nitrogen.

In the United States irrigated agriculture is a highly intensive food and fiber production system. Usually, the production per unit area is much greater in irrigated than rain-fed agriculture where crops are common to both. This may not always be the case, since the frequency and amount of precipitation in some rain-fed agricultural areas are sufficient that yield-limiting water stress may not develop. Also in some irrigated areas of the world, even though sufficient water is applied to avoid moisture stress of plants, other inputs may not be adequate to achieve high production per unit area. This is especially true in some of the less-developed countries where chemical fertilizers and pesticides have not been used to any great extent, even though there may be adequate water for irrigation.

IRRIGATED AREAS

Even with the limited production capacity of some irrigated areas, it is estimated that about 25 percent of the world food production is accomplished on the 15 percent of the cultivated lands that are irrigated. Table 4-1 shows the proportion of the total cultivated and irrigated areas for different regions of the world according to information available in 1988 (FAO, 1989). The table includes

Table 4-1.   Comparisons of total, cultivated, and irrigated land areas based on
               those countries within each of the regions having major irrigated areas
               in 1988.

Region	Total[a]	Cultivation[b]	Irrigated
	-------------------	hectares x 10⁶	------------------
Africa	2,963.6	185.4	11.0
Asia	2,678.2	450.9	142.0
Australia	761.8	47.1	1.8
Europe	473.0	140.1	16.8
North & Central America[c]	2,137.8	273.8	25.7
South America	1,753.5	142.0	8.6
Soviet Union	2,227.2	232.6	20.4
USA	916.6	189.9	18.1
TOTAL	12,998.1	1,471.9	226.3
World	13,076.5	1,473.7	227.0

Source: FAO (1989). Copyright © 1989 by Food and Agriculture Organization of the
United Nations. Reprinted by permission.

[a] Total land -- excludes inland water bodies (major lakes and streams).

[b] Arable and permanent crop lands -- land under temporary crops (double-cropped areas
counted only once) and lands with fruit trees, ornamental shrubs, nut trees and vines
but excluding trees grown for wood or timber.

[c] USA included.

only those countries that have some irrigated agriculture. Consequently, the total
cultivated area of the world would be greater if those countries without appreciable irrigation were included.

Another feature that makes irrigated agriculture important to world food production is that many of the crops that are used to provide a diverse and high-
quality diet are more adapted to the climatic conditions that prevail in the irrigated areas of the world. In addition, because of limitations of climate, topography, and high water-erosion potential, the most suitable land areas for rain-
fed agricultural production are already developed. In some developing countries
deforestation is being undertaken to provide more area suitable for rain-fed
agriculture. But mostly in developing countries, especially in arid regions, food
production for upgrading the diet and meeting increased demand will have to
occur where land resources have not been fully developed and where the potential exists for developing and using surface and groundwater supplies for irrigation.

In the United States most of the surface water supplies in the irrigated arid
and semi-arid West have been developed for irrigation purposes. In most of the
same areas underground water supplies have also been extensively developed.
In many of these areas the water table is dropping rapidly because of little or
no recharge. As a result, the cost of pumping the water is becoming too great
for growing low- and medium-income crops. These areas occur mainly in the
western United States where both seasonal and yearly rainfall are so low that
surface water catchment and groundwater recharge are not sufficient to offset
withdrawals. Many of these areas also have land potentially suitable for irrigation if additional water supplies become available.

Some distinction needs to be made between areas for which irrigation is obligatory for crop production and areas where supplemental irrigation is practiced primarily to enhance production. Where precipitation is less than about 20 to 25 cm per growing season, crop production under normal cultural practices tends not to be economically feasible. Under these conditions, moisture stress will be so severe that crops are unable to mature. In certain specialized cases, such as where clean fallow is practiced to allow maximum storage of water in the soil profile for more than one year, or where plants are widely spaced so that competition for the soil-water reservoir is minimized, it may be possible to grow crops in low-rainfall areas. However, the yield is usually so minimal that the effort is not made except in cases where food is in short supply. There are also seasonal periods of drought that make it essential to use irrigation for crop production. Areas of the United States where yearly or seasonal water deficiencies make irrigation essential are discussed in Chapter 6, Environment.

In other areas, irrigation may not be obligatory for crop production. Thus it may be considered supplemental, although necessary to produce the high yields essential in intensive agricultural production systems. It is in these areas of the United States that there has been expansion in irrigated acreage.

According to data in Table 4-2, slightly more than 95 percent of the irrigated land in the United States is found in the 17 western states plus Arkansas and Florida. The National Commission on Water Quality (1976) reported that every state in the United States had some irrigated area. However, areas where seasonal precipitation is limited are where irrigation is most extensively practiced. In the remainder of the United States, irrigation is principally utilized to augment normal rainfall to prevent periods of moisture stress during critical stages of development of shallow-rooted plants or plants growing on very sandy soils. Also shown in Table 4-2 are the percentages of irrigated areas in each state irrigated by surface, sprinkler, and drip methods, which will be discussed more fully in subsequent parts of this chapter.

The total amount of water used from surface or groundwater supplies for irrigation in selected states of the United States and the average amount of water withdrawn per irrigated hectare are shown in Table 4-3 for 1985. States where irrigation is mandatory for crop production have greater water diversions than those states in which irrigation is supplemental. A comparison of Arizona, New Mexico, and Washington with Arkansas, Kansas, and Nebraska illustrates the point. Another interesting comparison is the change in irrigated land area from 1975 to 1990, as indicated in Table 4-2. There was an increase of nearly 5,700,000 ha of irrigated land in the United States while there was a decrease of 833,000 ha in the 17 western states plus Arkansas and Florida. While large underground reservoirs of water and adequate surface supplies may be available for irrigation in the states not listed in Tables 4-2 or 4-3, the distribution of rainfall is adequate to provide water required by the plants grown. Where irrigation is practiced in the states not listed in Tables 4-2 and 4-3, it is assumed that irrigation is mostly utilized for shallow-rooted crops and/or for crops grown on very sandy soils with low water-holding capacity. In these cases, the frequency of precipitation may not be adequate to prevent moisture stress.

Table 4-2.   Irrigated land area and percentage irrigated by different methods for slected states.

State	Area Irrigated[a,b] ha x 1000 percent			Percent of irrigated area using Surface		Sprinkler		Drip	
	1975	1990	change[c]	1975	1990	1975	1990	1975	1990
Arizona	466	486	+4	95.4	85.0	4.2	12.5	0.4	2.5
Arkansas	687	758	+10	95.7	90.2	4.3	9.2	---	0.6
California	3,682	3,762	+2	81.4	62.6	16.2	32.2	1.4	5.2
Colorado	1,255	1,237	-1	81.9	71.7	18.1	28.3	---	---[d]
Florida	1,145	823	-28	38.8	26.4	60.7	51.5	0.5	22.1
Idaho	1,652	1,652	0	64.7	50.0	35.3	50.0	---	---
Kansas	1,175	1,175	-4	74.1	46.9	25.9	53.0	---	0.1
Montana	1,255	1,219	-3	91.7	67.5	8.3	32.5	---	---
Nebraska	2,550	3,364	+32	63.1	49.8	36.9	50.2	---	---
Nevada	586	529	-10	97.9	88.6	2.1	10.4	---	1.0
New Mexico	433	354	-18	85.1	64.1	14.8	35.4	0.1	0.5
North Dakota	42	77	+83	39.9	23.7	60.1	76.3	---	---
Oklahoma	381	249	-35	52.3	40.0	47.7	58.9	---	1.1
Oregon	792	750	-5	56.5	44.1	43.4	55.6	0.05	0.3
South Dakota	120	185	+54	19.6	22.1	80.4	77.9	---	---
Texas	3,318	2,539	-24	74.7	65.1	24.0	34.1	3.0	0.8
Utah	772	496	-36	79.1	78.4	20.9	21.6	---	---
Washington	653	802	+23	48.8	22.3	51.1	76.2	---	1.5
Wyoming	741	671	-9	91.0	90.0	90.0	10.0	---	---
Total	21,957	21,957[e]	-4						

Total for all other States	1,074	2,598	+41
Total for the United States	23,031	23,722	+3

Irrigation Method	Irrigated Area in hectares 1975	1990	% change
Surface	16,290,978	12,453,942	-30.8
Sprinkler	5,595,449	8,228,765	+47
Drip	70,573	141,293	+52.5

Source: Anonymous, 1975 & 1991.

[a] Area values are rounded to nearest 1000 hectares.

[b] Sub-surface irrigated areas are generally a small percent and rarely reported. In California, approximately one percent of the land is sub-irrigated (Stewart, 1975).

[c] Percent change between 1975 and 1990. Data from Irrigation Journal, 1975 & 1991 as cited in source above.

[d] Dashed lines indicates none or less than 200 hectares under drip irrigation.

[e] Total value differs from the total of the column due to rounding errors.

FACTORS INFLUENCING IRRIGATION PRACTICES

The amount of water actually applied to crops varies with (1) factors affecting crop water requirement and (2) factors influencing the efficiency of irrigation (Stewart, 1975). Of those factors that affect crop water requirements, climate is generally the most dominant. As has already been indicated, rainfall frequency, amount, and distribution determine those areas where irrigation is needed to augment precipitation or where irrigation is essential to produce a crop. Another important aspect of climate is the effect of temperature on evaporation of water from soil and plant surfaces and transpiration (evapotranspiration), which is generally referred to as the consumptive use of water.

Other factors that influence water requirements are the crop type and variety within a given plant species. Physical characteristics of plants that regulate the rate of water loss by plants include the number of stomatal openings in the leaves; and waxy, suberized, or heavily cutinized leaf surfaces. Other plant-related factors that influence crop water requirements include the length of the growing seasons, planting date, and rate of leaf canopy development to intercept

Table 4-3. Water used for irrigation in 1985 in selected states.

Selected states	Area irrigated ha x 1000	Water Withdrawn Surface	Groundwater	Amount per hectare ha-m
		ha-m x 1000		
Arizona	533	418	346	1.43
Arkansas	819	75	457	0.65
California	1,880	2,801	1,419	1.09
Colorado	1,359	1,419	295	1.26
Florida	775	184	220	0.52
Hawaii	107	79	47	1.17
Idaho	1,661	2,394	458	1.72
Kansas	1,195	36	619	0.55
Louisiana	355	107	98	0.58
Mississippi	301	22	100	0.41
Montana	932	1,138	11	1.23
Nebraska	3,029	289	716	0.33
Nevada	314	359	104	1.36
New Mexico	382	228	162	1.02
North Dakota	83	12	9	0.26
Oklahoma	181	9	53	0.34
Oregon	827	724	65	0.96
South Dakota	83	48	16	0.40
Texas	2,734	373	749	0.41
Utah	450	443	56	1.11
Washington	656	597	87	1.04
Wyoming	733	740	41	1.07
TOTAL	21,939	12,516	6,181	Avg. 0.83
TOTAL FOR OTHER STATES	1,227	70	137	Avg. 0.17
TOTAL FOR THE UNITED STATES	23,166	12,587	6,318	Avg. 0.82

Source: Adapted from Solley, Merk, and Pierce, 1988.

solar radiation and provide complete ground cover. Frequently, seeding rate and plant spacing will have an impact on water use by affecting how rapidly the soil surface is covered by the plant canopy. In general, peak transpiration of water is associated with maximum surface area of leaf cover and highest ambient air temperatures.

Another set of factors affecting the amount of water applied to crops are the soil physical characteristics. They influence the ability to replenish the water reservoir in the soil profile and the water-holding capacity. Irrigation management practices such as frequency, length of irrigation, and the amount of water applied are still other factors that influence the efficiency of irrigation.

IRRIGATION EFFICIENCY AND WATER CONSERVATION

Because of widespread public concern over the use of water and development of water supplies for irrigation, there is a great effort being made in conservation and efficient use of irrigation water. In an attempt to provide some guidelines for determining what constitutes a reasonable use of irrigation water, the California Department of Water Resources (1976) developed guidelines whereby the reasonable or unreasonable use of irrigation water could be evaluated. Within the guidelines in Table 4-4 are concepts of reasonable and unreasonable use of irrigation water that will impact the irrigation methods that may be used as well as other aspects of irrigation management.

METHODS

In general, the methods of irrigation can be classified into three categories. These are gravity (surface and subsurface) irrigation, sprinkler irrigation, and drip irrigation. As seen in Table 4-2, the vast majority of the irrigated land area in the United States is irrigated by some method of surface irrigation. Usually only small areas of subsurface irrigation are involved so this is not reported separately. In some isolated regions it may be the predominant method of irrigation. Sub-irrigation can only be practiced in areas where water tables are sufficiently high that the water table can be raised close to the surface so that capillary rise can provide unsaturated flow of water upward and replenish the soil-water supply. Once the supply is replenished, the water level in the supply ditches is dropped to allow drainage of the root zone, permitting aeration and root development.

Surface

There are many variations of surface irrigation used in different parts of the world, but conceptually they may be classified into three general types. One is flooding between borders. This method utilizes strips of varying lengths and widths depending upon the slope and soil texture. A frequently practiced variation of border flooding is to have the borders follow contour lines on approximately 1-ft (30-cm) elevation differences. This technique is generally practiced on relatively shallow soils or soils that are otherwise not particularly suited to the extent of land leveling required in order to use the rectangular border strip method of irrigation.

A second surface irrigation technique is basin flooding. In this technique, relatively small basins are constructed in the field rather than having a strip run the entire length of the field. This permits water to be applied to an individual basin either from an adjacent basin or from a supply ditch that allows each of the basins to be filled directly from it. With the advent of laser leveling, entire fields may be leveled to "zero" slope permitting the use of large basins comparable to border basins. In this case the soil texture and amount of water available

Table 4-4.   Concepts of reasonable and unreasonable use of irrigation water.

A. Reasonable Use

Reasonable uses of irrigation water under normal water supply conditions include the application of sufficient water to provide for the following:

1) The evapotranspiration (ET) requirement. Sufficient water must be provided to plants to meet their needs for evapotranspiration (ET) so that they will not be stressed, but will have an opportunity to maximize their use of solar energy in producing food, fiber, and/or other plant material.

Example: Fruit trees are particularly sensitive to water deficits during fruit development periods. Water stress at that time results in less production and also affects buds for the next crop.

2) The leaching requirement. In addition to the ET requirement, sufficient water must pass through the soil to prevent accumulation of harmful amounts of soluble salts in the root zone. Best current technology for determining and applying the leaching requirement should be demonstrated.

Example: Assume that the irrigation water contains 1,000 ppm salt concentration, and that the growing plants would be damaged by a concentration of more than 5,000 ppm in the soil solution at the bottom of the root zone. In this case, one-fifth of the irrigation water must be allowed to pass below the root zone in order to remove enough salts so the plant's tolerance would not be exceeded. This would be a leaching requirement of 20 percent.

3) The germination and emergence requirement: Seeds and seedlings may be inhibited by high temperature or by salt in the seed-bed that would not affect the fully developed plant. Emergence of seedlings may be blocked by a soil crust. Irrigation to correct these problems may be required in addition to normal evapotranspiration and leaching requirements.

4) Practical levels of irrigation efficiency: Applications of sufficient water using effective irrigation systems and prudent management to provide adequate soil moisture and meet the leaching requirement is necessary. Even with good management the best irrigation systems are less than 100 percent efficient.˙ Because the systems may not apply the same amount of water to every square foot of land and water infiltrates different soils within a field at different rates more water may have to be applied than is theoretically required to meet the leaching requirement in all parts of a field. Where infiltration rates vary widely within a field, some salt buildup and lower yields in the tighter soils may be necessary to prevent using excessive quantities of water for leaching.

5) Crop protection: This includes application of water to plants for frost protection, crop cooling, or pest management. This use may or may not be in excess of other basic requirements.

6) Other reasonable uses of irrigation water: In addition, it is reasonable to apply water in excess of the forgoing needs where the additional water is merely retarded in its flow downstream or stored underground, if it is later beneficially used, there is no significant harm to in-stream needs, no significant evapotranspiration or net energy loss, and where the quality of runoff or percolating water is not significantly degraded. This includes deliberate application of water to land beyond current needs, so that water can be stored underground for future recovery or for replenishment of groundwater overdraft if this is a reasonably efficient method of recharge.

There may be reasonable uses of irrigation water not listed above. Reasonableness of use should be considered on a case-by-case basis.

B: Unreasonable Uses

Unreasonable uses of water are those in excess of reasonable uses (described above) which lead to waste or degradation of water. These include:

Continued--.

72

Table 4-4 (continued)

1) Application of excess irrigation water because of low irrigation efficiency that could reasonably be corrected unless the excess can effectively serve downstream diversion or in-stream needs in a reasonable manner without significant water or energy loss or environmental damage.

Examples:

(a) Field runoff resulting from poor land grading, less control during irrigation than expected of a prudent farmer or lack of "tailwater" recovery by the user or by the water supply agency.

(b) Excess application and/or mal-distribution of water through irrigation systems because of poor design, mechanical defects, worn or improper equipment, or incorrect operation.

(c) Loss and mal-distribution of water caused by unnecessary operation of sprinklers under unfavorable conditions such as high winds.

(d) Irrigation in excess of need because of avoidable conflicts with other farming operations.

(e) Irrigation in excess of need because of unnecessarily rigid delivery schedules set by the water supply agency which do not allow the on-farm flexibility in water application needed for efficient operation.

2) Application of irrigation water in excess of plant growth and other requirements where the excess contributes unnecessarily to the quality degradation of underground or downstream waters.

Example: Excessive irrigation of plants that have recently received high levels of nitrogen fertilization.

3) Discarding of water of suitable irrigation quality in a manner that prevents further use.

Examples:

(a) Usable surface runoff that flows into a pond or other places with nearly impermeable bottoms and is allowed to evaporate.

(b) Drainage water of suitable irrigation quality that flows into a drainage canal from which it is not recovered.

(c) Runoff or drainage water of suitable irrigation quality that enters an estuary, the ocean or some other body of salty water, unless the runoff or drainage water efficiently serves some beneficial purpose.

4) Seepage and leakage losses from canals, irrigation ditches and pipelines due to design, construction or maintenance below generally accepted standards** --unless the water is recoverable in an efficient manner or efficiently serves other beneficial uses.

5) Failure to use cultural practices that meet generally accepted standards for prudent farmers aimed at achieving reasonable yields per unit of water. This may include improper management of such procedures as fertilization, pest and weed control; or inefficient timing of farm operations which interact with irrigation.

Continued--.

Table 4-4 (continued)

Examples:

(a)   Excessively weedy fields which produce lower yields while consuming as much water as well managed fields.

(b)   Under-irrigation or poorly timed irrigation which reduces yields, thereby resulting in appreciably less plant production per unit of water applied.

6) Use of good quality irrigation water to dilute a degraded supply which otherwise would be unsuitable for plant production -- unless the mixed waters produce more yield than would the good quality supply alone.

Example: If water too saline for irrigation is diluted with good water, the mixture may be suitable for irrigation.   However, the leaching requirement will be increased -- possibly so much that the blended waters will produce no more yield than would have been obtained by using only good quality water and discarding the bad.

Source: California Department of Water Resources, 1976.

* Irrigation application efficiency is defined as the percentage of the applied water that is stored in the soil and is available for use by the growing plants. Thus, if one and a quarter inches are applied and one inch remains in the root zone, then the irrigation efficiency is 80 percent. However, the portion that drains below the root zone may be useful in satisfying a leaching requirement.

** Where examination shows that low water price has resulted in significantly lower "generally acceptable standards" in an area than commonly found elsewhere, higher standards will be used to judge reasonableness of water use.

determine the size of the basin. In these irrigation methods, the entire land surface is wetted at each irrigation.

A third technique for surface application of water is the use of furrows or corrugations (shallow furrows) for containing and controlling the movement of water through the field. Only a portion of the surface is wetted. Because of the greater control over the flow of water, this technique can be adapted to a greater variation in both downfield and lateral slopes. Various techniques have been employed to control stream size to avoid excessive erosion in fields with slopes greater than about one percent, and to allow for adjustments of stream size to control uniformity of application. Furrow irrigation is frequently practiced on laser-leveled fields for row crops.

Sprinkler

In situations where soils have an undulating topography and are otherwise not suited to the degree of leveling required for the use of flood or furrow irrigation, then sprinkler irrigation is generally used. Other factors such as nontillage, coarse textured soil, nonuniformity of soils, limited quantities of water, expensive water, or special uses such as frost control and seed germination may also favor the use of sprinklers for irrigation. Because of some of the varied conditions indicated, states such as Washington, South Dakota, Oregon, Oklahoma, North Dakota, and Florida have very large percentages of their total acreage under sprinkler irrigation. As with any irrigation system, numerous variations and modifications have been employed to adapt the technique to a particular location.

One technique is to use a portable set of sprinklers that can be moved manually from one area to the next as an irrigation is completed. Another technique is to use a mechanical method of moving the sprinklers. One such system is the so-called side-roll or wheel-line sprinklers in which some type of power unit moves the system from one point in the field to the next as an irrigation set is completed. A slight variation is to drag a sprinkler line behind the mechanically moved system, which allows a much wider irrigation set. Another mechanically moved system is the so-called center pivot. In this technique of irrigation, the sprinkler line is fixed at one end and then pivots around that point providing a circular irrigation pattern. One disadvantage of this technique is that all of a normally square or rectangular field cannot be irrigated. Cornering systems make it possible to irrigate about 95 percent of such fields.

Another sprinkler irrigation technique is the hose-drag system. This can also be a mechanically moved system, but it employs a flexible hose rather than a rigid pipe for water delivery. There is a relatively high labor requirement required to reset these systems, even though some portion of the movement is done mechanically.

The permanent set sprinkler system is generally limited to high-cash-value crops because of its high investment cost. Supply and lateral lines may be buried or left on the surface. The entire surface area can be irrigated without moving the system by sequencing the water over the field. This type of system is also well adapted to soils that have relatively low water intake rates since it is readily amenable to designs permitting relatively low application rates. An additional advantage of permanent set sprinkler systems is their use for frost protection and controlling other microclimate factors such as high temperature.

Drip

The drip system employs a network of plastic tubing for water delivery from the supply line under low pressure to the plants. Emitters are located at various points along the tubing, which reduce the energy of the water flow to a relatively low level by means of controlling pressure and outflow with a small-diameter orifice or a long flow path. The rate of water application generally ranges from 2 to 8 liters per hour, which allows water to move through the soil by unsaturated flow. Because of the relatively low rate of application, frequent applications of water may be necessary to meet the consumptive needs of the plant. The number of emitters can be adjusted to achieve an adequate supply of water to the plants. Because of reduction in the amount of surface area wetted and more precise control of the amount of water applied, an improvement in irrigation efficiency may result. This reduction of surface evaporation losses is greatest for orchard, vineyard, or annual crops where the crop canopy covers a small percentage of the surface. Another advantage of drip irrigation systems is their relatively low rate of application, which may be especially adapted to soils with extremely low infiltration rates.

SELECTION OF IRRIGATION METHODS

Some general guidelines that may be useful in determining the type of irrigation system most adapted to a particular set of conditions are outlined in Table 4-5. These factors must be considered in selecting an irrigation system or determining the suitability of a system to a particular set of conditions. In Table 4-6 are given some common problems associated with each of the irrigation systems and different management techniques that may be used to mediate or eliminate such problems.

Methods of Measuring Efficiency of Water Use

An important means for evaluating any irrigation system is the water application efficiency. This is determined by the degree of water replenishment in the root zone at each irrigation, the amount of runoff or deep percolation, and how uniformly the water was distributed during an irrigation. There are different concepts of efficiency used to evaluate the efficiency of water use. For example, Israelsen and Hansen (1967) presented methods for calculating water conveyance efficiency, water application efficiency, water use efficiency, water storage efficiency, water distribution efficiency, and consumptive use efficiency. Consequently, referring to efficiency of water use or irrigation efficiency without being specific as to which of these terms is involved does not communicate the concept adequately. At the same time, it is generally recognized that reference to irrigation efficiency is construed as meaning water application efficiency. Israelsen and Hansen (1967) also discuss a more complete method of evaluating the efficiency of an irrigation system by using the combination of water application, water storage, and water distribution efficiencies simultaneously. *Water application* efficiency (in percent) refers to the quantity of water stored in the soil root zone divided by the quantity of water delivered to the field. *Water storage* efficiency (in percent) refers to the quantity of water stored in the root zone during an irrigation divided by the amount of water needed in the root zone prior to an irrigation. The *water distribution* efficiency (in percent) is equivalent to the average absolute deviation in depth of water stored from the mean depth of water stored during an irrigation divided by the mean depth of water stored during the irrigation. For the water application efficiency, losses of water from surface runoff or deep percolation below the root zone reduce the water application efficiency. If evaporation loss is great, it will also reduce the water application efficiency.

In irrigated agriculture the application of water to replenish the soil reservoir is made before the soil reservoir is completely depleted. Thus, soils only need to have a portion of their available water-holding capacities replenished during an irrigation. Consequently, the water storage efficiency is an important criterion for evaluating irrigation efficiency. Because of scarcity of water, poor penetration or inproper application, the amount of water applied may not replenish the water in the root zone adequately to meet the water demands of the plant. Furthermore, in areas where the quality of the irrigation water is poor, low water

Table 4-5. Factors to consider when selecting an irrigation system.

Factors to consider	Sprinkler systems					Surface flood systems			Drip systems
	Portable	Wheel roll	Solid set	Center pivot	Boom (giant)	Graded border	Level border	Furrow	
Slope limitations									
Direction of irrigation	20%	15%	none	15%	5%	0.5-4.0%	Level	3%	none
Cross-slope	20%	15%	none	15%	5%	0.2%	0.2%	9%	none
Soil limitations									
Intake rate (in/hr) Min.	0.1	0.1	0.05	0.30	0.30	0.30	0.1	0.1	0.02
Max.	6.0	6.0	6.0	6.0	6.0	2.0	2.0	3.0	none
Water holding capacity in root zone	3.0	3.0	none	2.0	2.0	6.0	6.0	4.0	none
Depth	none	none	none	none	none	Soil should be deep enough to allow for grading required.			none
Erosion hazard	slight	slight	slight	slight	slight	moderate	slight	severe	none
Saline-alkali soils	slight	slight	slight	slight	slight	moderate	slight	severe	moderate
Water limitations									
Quantity									
Total dissolved solids (TDS)	severe	severe	severe	severe	severe	slight	slight	moderate	slight
Suspended solids	moderate	moderate	moderate	moderate	moderate	none	none	none	severe
Rate of flow	low	low	low	high	high	moderate	moderate	moderate	low

Continued--

Table 4-5 (continued)

Factors to consider	Sprinkler systems					Surface flood systems			Drip systems
	Portable	Wheel roll	Solid set	Center pivot	Boom (giant)	Graded border	Level border	Furrow	
Climatic factors									
Temperature Control	no	no	yes	no	no	yes	yes	yes	no
Wind Affected	yes	yes	yes	yes	yes	no	no	no	no
Adaptability To all crops[a]	good	good	good	fair	limited	very good	very good	very good	good
System costs 1976 data[a]									
Capital cost ($/acre)	400-600	400-600	700-1200	700-1000	600-700	500-600	500-600	400-500	500-1000
Labor costs[b]	high	moderate	low	low	moderate	moderate	moderate	high	low
Power cost[c]	high	high	high	high	high	low	low	low	moderate
Avg. annual costs[d] ($/ac/yr)	100-200	100-200	200-300	200-300	200-300	100-200	100-200	200-300	200-300
Application efficiency[e]	70-85	70-85	75-90	70-85	65-80	70-85	75-90	70-85	80-90

Source: California Interagency Agricultural Information Task Force, 1977a.

a The agricultural input cost index for 1989 is 101 with a 1977 base equal to 100. [Economic Indicators of the Farm Sector: Production and Efficiency Statistics, 1989. (1990)].

b Low = less than $20/ac/yr; Moderate = $20-50/ac/yr; High = over $15/ac/yr.

c Low = $0-5/ac/yr; Moderate = $5-15/ac/yr; High = over $15/ac/yr.

d Amortized capital cost plus operational and maintenance cost.

e Assuming good to excellent management.

Table 4-6.   Solution to common irrigation problems.

Problem	Sprinkler systems	Surface flood systems — Level borders (basin)	Graded border	Furrow	Drip systems
(1) Surface runoff	Decrease application by: A) replace worn nozzles, B) reduce nozzle size, C) reduce pressure, D) decrease set time, E) repair system leaks Increase Intake Rate by: A) cover crop in orchard B) surface mulching C) apply soil amendments	Repair border leaks	--------Decrease stream size--------		Repair system leaks
			--------Install tailwater return system--------		
			--------Increase run length--------		
			--------Decrease slopes--------		
			--------Increase intake rate by:--------		
			A) Cover crop in orchard		
			B) Apply soil amendments		
(2) Ponding at lower end	See (1) above	Re-level basin	--------See (1) Above--------		See (1) above
			--------Install tailwater return system--------		
(3) Slow advance of water stream	Not applicable	Not applicable	--------Increase stream size--------		Not applicable
			--------Decrease border width--------	Increase slope	
			--------Decrease border length--------	Decrease slope	
(4) Rapid advance of water stream	Not applicable	Not applicable	--------Decrease stream size--------		Not applicable
			Increase border width	Increase furrow length	
			Increase border length	Decrease slope	
			Decrease slope		

Continued--.

Table 4-6 (continued)

		SOLUTIONS			
			Surface flood systems		
Problem	Sprinkler systems	Level borders (basin)	Graded border	Furrow	Drip systems
(5) Uneven distribution	Check sprinklers for proper operation, repair leaks. Check operating pressure for pump and system shutdown during high wind. Change: A) Sprinkler spacing B) Sprinkler head C) Nozzle size	Adjust basin boundary to soil type	Regrade---------- Increase stream size-------- Use border checks ------Install tailwater return system-----		Check proper operating pressure; Check for clogged emitters; Change emitter spacing and/or location
(6) Erosion	See (1) and (6) above	Decrease inlet velocity	Decrease inlet velocity; Decrease stream size; Decrease cross slopes	Decrease stream size; Decrease slope; Change furrow slope	See (1) Above
(7) Muddy tailwater	See (1) and (6) above	Not applicable	See (6) above	See (6) above	Not applicable
(8) Saline water supply	Replace system with a more corrosion resistant material. Irrigate at night to reduce leaf burn.	----Maintain high soil moisture content by irrigating more frequently------ ----Apply adequate water for leaching----		Use special seedbed shapes for salinity control.	----------

Continued--.

Table 4-6 (continued)

Problems	Sprinkler systems	Solutions			
		Surface flood systems			Drip systems
		Level borders (basin)	Graded border	Furrow	
(9) Salt deposits on plants	Irrigate at night to reduce leaf burn.	Not applicable	Not applicable	Not applicable	Not applicable
(10) Salt deposits on soil surface	--------	--------Change to more salt tolerant crop--------			
	--------	--------Apply adequate water for leaching--------			Apply adequate water for leaching. Use other irrigation methods.
	--------	--------Improve subsurface drainage--------		Use special seedbed shapes for salinity control	
(11) High water table	--------	--------Improve water management--------			--------
	--------	--------Improve subsurface drainage--------			--------
(12) Uneven crop growth	--------	--------See (2), (5), (10), and (11) Above--------			--------
(13) Distribution system losses	Repair pipeline leaks	--------Replace open ditches with pipeline--------			Repair pipeline leaks
		--------Line open ditches or ponds with concrete, plastic, bentonite or chemical sealers--------			
		--------Install interceptor subsurface drains parallel to canal, ditch, or pond--------			

Source: California Interagency Agricultural Information Task Force, 1977b.

storage efficiencies may lead to salt buildup, which creates an even greater probability of plant water stress.

A final characteristic frequently employed in evaluating irrigation efficiency is the water distribution efficiency. According to Israelsen and Hansen (1967), the water distribution efficiency is identical to the formula for calculating the coefficient of uniformity that was developed by J.E. Christiansen in 1942. Evenness of distribution is a desirable characteristic of any irrigation system. If an irrigation applies water unevenly, excess water is applied in other parts of the field. Conversely, if adequate water is applied only to parts of the field, then water stress may develop elsewhere in the field. This has a tendency to cause yield reduction as a result of salt accumulation and water stress. It should be apparent that more than one parameter needs to be considered in evaluating irrigation efficiency and for comparing relative efficiencies of different irrigation systems.

Reported Water Application Efficiencies

The fact remains that water application efficiency is the factor that is predominantly used to compare irrigation systems. For example, in Table 4-5 water application efficiencies assuming good to excellent management are given for each of the irrigation systems. From this it can be seen there are relatively small differences in water application efficiency for properly managed systems. However, as pointed out by Reed et al. (1977), common irrigation (water application) efficiencies vary from 40 to 90 percent. The variation is dependent upon such factors as water quality, soil type, the length of irrigation run, slope, temperature, wind, and type of crop as well as method of irrigation.

Another means of evaluating irrigation efficiency is to determine the percent of water applied to the field that is used for evapotranspiration. This is water use efficiency as outlined by Israelsen and Hansen (1967), which is the amount of water beneficially used divided by the amount of water delivered. For most practical purposes, water use efficiency and water application efficiency may be considered equivalent. However, since the amount of water stored in the soil root zone may not all be evaporated or transpired, the application efficiency will tend to be somewhat greater than the water use efficiency. Irrigation efficiencies (water use efficiency) are reported in Tables 4-7 and 4-8 for different locations and a variety of crops. The Yuma Valley soils generally have a higher irrigation efficiency than those found in the Yuma Mesa area. One factor contributing to the decreased irrigation efficiency in the Yuma Mesa area is the sandy soils. Consequently, larger amounts of water are leached beyond the root zone during an irrigation. In the Yuma Valley, the soils are fine textured (silt loams to clay loams), resulting in the generally higher irrigation efficiencies. Low irrigation efficiencies are also attributable to shallow-rooted crops and the common practice of continuous irrigation during the period of seed germination and emergence for certain vegetables in order to prevent crusting of the soil surface from causing a severe stand reduction.

In Table 4-8 are reported irrigation efficiencies for various crops in the Sacramento and San Joaquin Valleys of California. The range of irrigation efficiencies

Table 4-7.   Irrigation efficiencies for various crops in the Welton-Mohawk irrigation and drainage district in Arizona.

| Crop | ------------Irrigation Efficiency[a,b]------------ | |
	Yuma Valley %	Yuma Mesa %
Alfalfa	82	55
Bermuda seed	66	--[c]
Citrus	--	28
Cotton	79	--
Lettuce	20	--
Melons	32	--
Pasture	60	36
Sorghum	56	--
Wheat	64	--

Source: Krull and Clark, 1977.

[a] The percent of water applied to the field that is evapotranspired.

[b] These values are three-year averages for surface irrigation methods for the district: 1970, 1971, and 1972.

[c] No values reported.

(water use efficiencies) does not seem too different from those reported for Arizona in Table 4-7, where only surface methods of irrigation were considered. The reported irrigation efficiencies in Table 4-8 are the averages for crops regardless of irrigation methods. However, approximately 62 percent of all the area is irrigated by surface methods (Anonymous, 1991). Consequently, the averages reflect to some extent the irrigation efficiency that may be achieved from surface irrigated methods. As shown in Table 4-2, in California use of surface irrigation has declined substantially during the past 15 years while the deciduous orchards, subtropical orchards, and vineyards are increasingly being sprinkler irrigated. For some of the higher-cash-value vegetable crops, such as tomatoes, increasing areas are being sprinkler irrigated or are using sprinklers for germination and stand establishment and then furrow irrigation to supply the water needs for the remainder of the season.

One might expect that application efficiencies as reported in Table 4-9 would be slightly greater than the water use efficiencies reported in Tables 4-7 and 4-8, as previously discussed, assuming the measurements were made under essentially the same conditions. However, these irrigation efficiencies show the same wide differences for different crops and these are at least as significant as the regional differences due to soil and climate.

The on-farm efficiencies reported in Tables 4-7, 4-8, and 4-9 may not be totally indicative of the overall project efficiency. As pointed out by Tanji et al. (1976), project efficiency may be quite high as a result of reuse of irrigation return flows. Tanji cited a California Department of Water Resources study on irrigation efficiency in the western part of the Sacramento Valley that showed that even with low on-farm efficiencies the downstream reuse of return flows

Table 4-8   Reported irrigation efficiencies* for various crops grown in the central valley of California.

Crop	Sacramento Valley		San Joaquin Valley	
	SWRCB[b]	DWR[c]	SWRCB	DWR
Alfalfa	68	68	61	66
Barley	-[d]	-	-	64
Beans (dry)	55	58	52	50
Cantaloupes	--[e]	--	-	32
Corn (field)	72	95	63	63
Cotton	--	--	66	68
Deciduous Orchard	76	73	67	80
Almonds	55	-	58	-
Grain sorghum	70	74	68	68
Pasture	57	57	60	55
Potatoes	-	-	32	-
Rice	40	37	52	52
Subtropical Orchard	57	-	83	-
Sugar beets	57	55	69	64
Tomatoes	65	71	96	71
Vineyard	57	57	60	62

Source: Tanji et al., 1976.

* Irrigation efficiency = $\dfrac{\text{Evapotranspiration of applied water}}{\text{Amount of applied water}} \times 100$

[b] SWRCB refers to the California State Water Resources Control Board.

[c] DWR refers to the California Department of Water Resources.

[d] Single dash line indicates no values reported.

[e] Double dash line indicates the crop is not grown in that region.

resulted in a basin efficiency of approximately 70 percent for the 1960 to 1970 period. Furthermore, he found the irrigation efficiency to be increasing with time due to greater reuse as a result of slight increases in irrigated area without a concomitant increase of the water supply.

Runoff

One of the contributing factors to low water use efficiency is the amount of runoff. In general, any attempt to provide greater control of the water during its application such as by use of basins, furrows with tailwater returns, or sprinklers results in a considerable reduction in the percent of applied water that runs off the field. While the estimated percentages of runoff water in Table 4-9 are not meant to be representative of all of the irrigated areas of the western United

Table 4-9.   Estimated irrigation efficiencies, nitrogen leaching losses and sediment yields by crop for different surface irrigation systems[a].

Crop	Furrow			Furrow with cut-back		
	Irrigation efficiency[b]	Nitrogen leaching loss[c]	Sediment yield	Irrigation efficiency	Nitrogen leaching loss	Sediment yield
	%	kg\ha	mt\ha	%	kg\ha	mt\ha
Potatoes	40.5	60.3	20.0	57.7	46.8	8.3
Sugar beets	49.9	55.7	16.2	66.0	40.4	5.8
Wheat	46.3	19.6	0.7	63.3	15.7	0.2
Spearmint	40.5	61.4	2.5	57.7	42.8	0.9
Alfalfa	57.9	--[d]	1.4	73.4	--	0.5

Source: Gossett and Whittlesey, 1976.

[a] Irrigation systems had no tailwater reuse systems.

[b] Irrigation efficiency is defined as the percent of water applied to the field that is stored in the root zone.

[c] The amount of nitrogen applied for which leaching losses were calculated are: potatoes, 224 kg/ha for both irrigation systems: sugar beets, 196 kg/ha and 198 kg/ha for the furrow and furrow cut-back systems, respectively: wheat, 140 kg/ha for both irrigation systems: spearmint, 196 and 168 kg/ha for furrow and furrow cut-back systems, respectively.

[d] No nitrogen was applied.

States, they do provide an indication of the order of magnitude of surface runoff under a rather wide range of conditions for each of the indicated irrigation methods. Comparison of these runoff values can be made by reconciling the irrigation efficiencies in Table 4-10 with the irrigation efficiencies in Tables 4-7 and 4-8. The efficiency of furrow irrigation as shown in Table 4-9 was

Table 4-10.   Water-use efficiency, percent surface run-off and sediment yield for various crops in Boise Valley, Idaho.

Crop	Water-use[a] Efficiency ----%----	Surface Run-off ----%----	Sediment Yield -kg/ha-
Alfalfa	54	17	73
Barley	>100[b]	0	0
Dry beans	63	33	12,999
Hops	42	28	3,414
Lima beans (1972)	>100	28	2,972
Lima beans (1974)	>100	0	0
Onions (1971)	48	40	2,587
Onions (1972)	44	44	2,998
Popcorn	74	20	264
Seed corn	56	17	3,861
Sugar beets (1971)	63	32	1,057
Sugar beets (1972)	50	49	1,558

Source: Fitzsimmons et al., 1975.

[a] Water-use efficiency is the percent of applied water evapotranspired.

[b] Values greater than 100 percent imply there was net upward movement of groundwater into the root zone.

lower than with the furrow method using a cutback system (reduction of stream size once water has reached the end of the field). Lack of control of the irrigation stream resulted in greater runoff and thus a lower irrigation efficiency.

Irrigation efficiencies for similar irrigation methods across a wide range of conditions are fairly similar. The water use efficiency for a number of crops grown in Idaho reported in Table 4-10 are similar to the irrigation efficiencies reported elsewhere in this chapter. The reported range of surface runoff values was within the 0 to 50 percent range. One of the factors contributing to higher runoff and lower efficiencies is the greater slope on which irrigation is practiced in the Boise Valley as opposed to the Central Valley in California and most areas in Arizona. Irrigation efficiency parameters along with soil factors should be used to determine the impact of surface runoff on specific fields. In Table 4-10 the sediment yield on fields where dry beans were grown was substantially higher than for other crops. This reemphasizes the necessity of determining the characteristics of each field in order to properly evaluate the impact of irrigation on sediment yield or to be able to conclude that one crop may be more likely to contribute to sediment load than another. Without having specific information, one might conclude that the slope was greater in the dry bean field, thereby contributing a greater sediment loss for the same amount of water runoff.

With respect to sediment yields or losses as a function of irrigation, a rather interesting relationship was shown by Fitzsimmons et al. (1977) as in Fig. 4-1. As the water application efficiency increased for the field under investigation, the sediment loss from the field decreased in curvilinear fashion. At very low

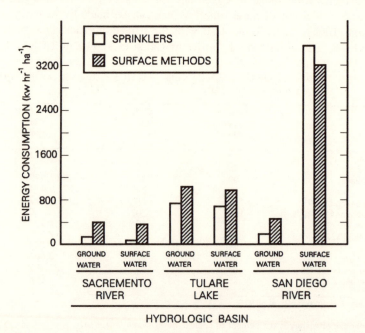

Figure 4-1   Sediment yield versus application efficiency for a bean field near Kimberly, Idaho, 1976. (Fitzsimmons et al., 1977.)

application efficiencies a small increase in efficiency can result in a large reduction in the amount of sediment loss. Conversely, at high water application efficiencies a large change in water application efficiency may only cause a small decrease in sediment loss. While this is not sufficient information upon which to base a principle, it does conform with what one might intuitively assume to be the case. Consequently, it may serve as an example that may be used as a preliminary guideline until demonstrated otherwise.

Such broad ranges in runoff as have been reported are not really informative. In isolated situations the percent runoff can even be greater. Therefore, estimates of actual runoff percentages will have to be made in the locale and more specifically for the field on which the effect of the runoff is being evaluated.

Costs

Whether contemplating the installation of a new irrigation system or replacement of a current irrigation system, some of the principal features to be considered are the initial capital costs, overhead, and operating costs. These factors determine the alternative irrigation systems available for an individual farm and given set of conditions, which in turn influence the water application efficiency. The relative comparisons of system costs in Table 4-11 are made without any reference to suitability or adaptability of the system to a particular set of conditions. The costs are meant to provide some indication as to the magnitude of costs and factors to consider when comparing systems. Although the costs are for the 1977 year, according to the U.S.D.A. Economic Research Service (1992) statistics for 1990 the agricultural inputs costs index is essentially the same as it was in the base year 1977. The information is presented for selected irrigation systems in ascending order of cost per hectare. The relative ranking may change as a function of land and power costs.

Land cost becomes a factor with the center pivot system since unused land as a result of inability to irrigate the total area will differ in value. The total cost of the unused land must be included in the cost of the system. Another variable cost is the cost of power for pumping and pressurizing sprinkler irrigation systems. With surcharges now being assigned to large power users and rapidly increasing energy cost for power development, the power cost for pressurizing certain systems may change the relative order of the irrigation systems listed in Table 4-11.

To further indicate the importance of the energy usage in relation to different irrigation methods, Fig. 4-2 shows the amount of energy required to irrigate alfalfa for different hydrologic basins. Knutsen et al. (1977) pointed out that sprinklers may reduce the amount of water needed, but they require additional energy per unit applied water to pressurize the system. They calculated the amount of energy to pressurize a typical sprinkler system at 55 psi (3.75 atm) would be 216 kW-hr per acre foot of water (1,752 kW-hr per hectare-meter). This assumed an overall pumping plant efficiency of approximately 60 percent. The energy requirement is greater for sprinkler irrigation methods as seen in Fig. 4-2 for the Sacramento and Tulare hydrologic basins whether the water supply is surface or groundwater sources. The same relationship holds true for

Table 4-11. Sample costs for several methods of irrigation[a].

Item	Flood		Center Pivot		Side roll or wheel line sprinkler	Hand move sprinkler	Furrow tailwater reuse	Hose drag	Permanent set sprinkler	Drip
	Surface water	Ground water	Corner device	No corner device						
Times irrigated	6	6	8	8	8	8	6	8	8	frequent
Total water applied										
(acre-feet/acre)	3.5	3.5	3	3	3	3	3.5	3	3	2
Investment per acre, US$[b]										
Grading & leveling	200	200					100			
Well	100	100	100	100	100	100	100	100	100	100
Pump	75	75	130	130	135	135	75	135	135	80
Pipeline & valves	250	250					250			
Sprinkler or drip system			325	270	285	225		360	1100	800
Tailwater reuse system	25	25					25			
TOTAL (investment)	475	650	599[c]	730[c]	520	460	300	595	1335	980
Overhead costs per acre[d]	50.25	69.59	81.70	91.50	71.45	61.85	32.00	92.45	146.85	147.80
Operating costs per acre[e]	67.10	44.20	44.20	42.00	69.15	102.75	153.75	111.75	72.75	78.15
TOTAL (irrigation costs per acre)	117.35	123.79	125.90	133.50	140.60	164.60	185.75	204.20	219.60	225.95

Source: Adapted from Reed et al., 1977.

[a] Calculations for center pivot irrigation are based in 160 acres (65 ha), all other irrigation system costs are based on 100 acres (40 ha).

[b] The Agricultural Input Cost Index for 1990 is 101 with a 1977 base of 100 (USDA Economic Research Service, 1992).

[c] Values differ from the sum of costs in the columns because of costs assigned to unused land: $48.40 for the cornering system and $253.00 for the system without the cornering capability.

[d] Overhead costs include depreciation, interest, and taxes on the investment.

[e] Operating costs include land preparation for irrigation, labor, water, power and system maintenance. Power costs for these calculations was approximately $0.03 per kilowatt-hour.

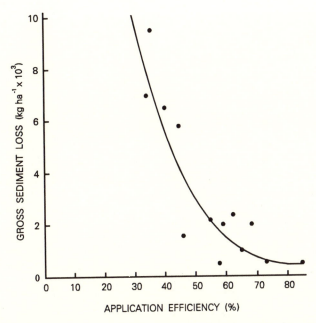

Figure 4-2 Energy required to irrigate alfalfa in three hydrologic basins of California, comparing ground water versus surface water and sprinkler versus surface methods assuming 20 percent less water use by sprinklers. (Knutson et al., 1977.)

groundwater sources in the San Diego hydrologic basin; however, for the surface water, the surface irrigation methods have higher energy requirements. This is attributed to the large amount of energy required to pump Colorado River water into the San Diego River basin even though it was assumed there would be a 20 percent savings of water attributed to the use of sprinkler irrigation. This emphasizes how important it is to consider the energy cost associated with different irrigation systems and water resources. It also indicates that maintaining the water pumping plant in efficient operating condition is essential to avoid unnecessary power costs.

Water Movement and Distribution

As a result of the uniformity of application and application efficiency associated with different irrigation systems, the movement and distribution of water can be quite different from one system to the next. These differences are coupled with the fact that nitrate may move out of the root zone and potentially contribute to pollution of receiving waters.

As pointed out earlier in this chapter, the efficiency of water application and water storage in the root zone is determined to some extent by the irrigation method. It is also dependent upon the amount of water that needs to be replenished within the soil profile. In Table 4-12 some guides are given for determining the amount of water that has been depleted from the soil using a feel method. As crude as this technique may be, it is a commonly used method for estimating

Table 4-12. Guide for estimating available soil moisture depleted by the "feel" method.

Soil moisture deficiency	Feel or appearance of soil and moisture deficiency in inches of water per foot of soil			
	Coarse texture (loamy sand)	Moderately coarse (sandy loam)	Medium texture (loam)	Fine and very fine texture (clay loam, clay)
0% (field capacity)	Upon squeezing, no free water appears on soil but wet outline of ball is left on hand*. 0.0	Upon squeezing, no free water appears on soil but wet outline of ball is left on hand. 0.0	Upon squeezing, no free water appears on soil but wet outline of ball is left on hand. 0.0	Upon squeezing, no free water appears on soil but wet outline of ball is left on hand. 0.0
0 - 25%	Tends to stick together slightly, sometimes forms a very weak ball under pressure. 0.0 to 0.2	Forms weak ball, breaks easily, will not stick. 0.0 to 0.4	Forms weak ball, is very pliable, sticks readily if relatively high in clay content. 0.0 to 0.5	Easily ribbons out between fingers, has slick feeling. 0.0 to 0.06
25 - 50%	Appears to be dry, will not form a ball with pressure. 0.2 to 0.5	Tends to ball under pressure, but seldom holds together. 0.4 to 0.8	Forms a ball, somewhat plastic, will sometimes stick slightly with pressure. 0.5 to 1.0	Easily ribbons out between thumb and forefinger. 0.6 to 1.2
50 - 75%	Appears to be dry, will not form a ball with pressure. 0.5 to 0.8	Appears to be dry, will not form a ball. 0.8 to 1.2	Somewhat crumbly, but holds together from pressure 1.0 to 1.5	Somewhat pliable, will ball under pressure. 1.2 to 1.9
75 - 100% 100% is permanent wilting %	Dry, loose, single grained, flows through fingers. 0.8 to 1.0	Dry, flows through fingers. 1.2 to 2.0	Powdery, dry, sometimes slightly crusted but easily broken down into powdery condition. 1.5 to 2.0	Hard, baked, cracked, sometimes has loose crumbs on surface. 1.9 to 2.5

Source: Adapted from Israelsen and Hansen, 1967. Copyright © 1967 by John Wiley & Sons, Inc. Reprinted by permission

* Ball is formed by squeezing a handful of soil very firmly.

the soil-water deficit in the field. With experience this technique can be very useful in determining the appropriate time to irrigate when coupled with observation of plant symptoms. Furthermore, by utilization of an auger to examine the moisture content of the soil throughout the profile, one can roughly determine the effectiveness of an irrigation in replenishing the soil water in the rooting zone or the uniformity of water application. These are not the only techniques one can use to determine when to irrigate or the effectiveness of an irrigation. More sophisticated techniques such as tensiometers, gypsum blocks, neutron probes, infrared thermometry, and time-domain reflectometry are used and others are currently being developed. But these are not in common use by farmers and field representatives of agribusinesses or by consultants. A tool used increasingly frequently is evapotranspiration calculated from climatic variables using temperature, relative humidity, and solar radiation. This technique is gaining wider usage with installation of automated climate stations and networks. Meteorological networks are now operational in Arizona, California, and Nebraska.

Furrow Irrigation

The direction of water movement and its distribution under different irrigation systems has a definite impact on the amount of nitrate leaching that may occur. Figure 4-3 shows the relative direction and extent of water movement under a furrow irrigation system. The arrows underneath the furrow and in the center of the bed indicate the general direction of water movement. The space between the dotted lines is indicative of the infiltration rate: as the space narrows the infiltration rate is nearly steady. With time the surface of a bed may become wetted as a function of lateral movement or subbing of the water between the furrows. The arrows in the center of the bed down through the soil profile indicate the relative direction of water movement. This becomes important when considering the placement of nitrogen fertilizers within a bed. If fertilizer were

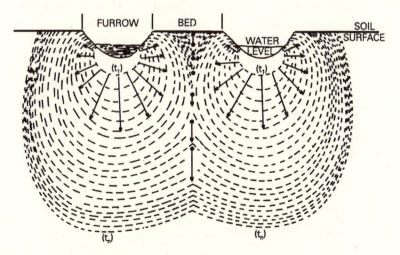

Figure 4-3 Diagram of direction and relative extent of water movement as a function of time (*t*) for furrow irrigation.

placed in the bed at the same depth as the bottom of the furrow, the direction of movement would be towards the center of the bed with little net movement either upward or downward. Slightly higher or lower placement may allow water to move the nitrogen either towards the surface of the bed or down through the soil profile, respectively. It is precisely this influence of water movement on nitrogen distribution that makes it necessary to understand how water moves in the soil profile. Figure 4-3 also indicates that rather shallow placement on the side of the bed or in the top of the bed would lead to nitrate accumulation near the surface of the bed. For all practical purposes, this removes the nitrogen from the root zone and makes it unavailable to the plant. Conversely, nitrogen placed in the bottom of the furrow has the greatest potential for nitrate leaching because in this zone all movement is downward.

Border Irrigation

While there is opportunity to place nitrogen in a furrow irrigation system in such a way as either to cause greater loss or enhance plant uptake, little opportunity exists for this kind of mechanical manipulation of placement of nitrogen in a border flood irrigation system (Fig. 4-4). Except for the edges of the flooded area in proximity to the borders all water movement is downward. Consequently, nitrate moves downward through the soil profile. There will be some tendency for nitrogen to accumulate in the borders, but the amount is small compared to the total area. Thus, most of the nitrogen under this situation is subject to the same leaching potential. The extent of leaching under these conditions is very much dependent upon the quantity of water applied and the application and storage efficiency of the irrigation.

Sprinkler Irrigation

The direction of water movement under sprinkler irrigation systems is also in the downward direction (Fig. 4-5). At the perimeter of the sprinkler irrigation set, there may be a small tendency for lateral movement, as indicated by the arrows at the extremes of the irrigation pattern. An interesting aspect of sprinkler irrigation in relation to border irrigation is the differences in the depth of water penetration as a function of the sprinkler pattern. In flood irrigation it was

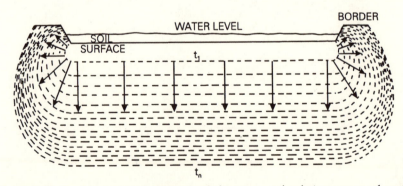

Figure 4-4 A cross-sectional diagram of direction and relative extent of water movement as a function of time (*t*) for flood irrigation.

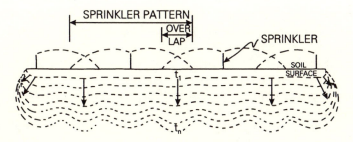

Figure 4-5 A cross-sectional diagram of direction and relative extent of water movement as a function of time (*t*) for sprinkler irrigation.

assumed that the water was uniformly distributed over the field at any given cross-section of the field. However, under sprinkler irrigation this may not be the case. Generally, these systems are designed to give as uniform as possible an application taking into account the spray distribution for a single sprinkler and the overlap pattern from adjacent sprinklers. Nevertheless, nonuniformity can occur during a particular irrigation and the wetting pattern may look similar to that indicated in Fig. 4-5. As with the flood irrigation, any placement of nitrogen fertilizer will be subjected to the same general downward movement. Consequently, little opportunity is available to influence downward movement by different fertilizer placement in the field. However, with sprinkler irrigation, the smaller amount of water per application is less apt to cause leaching of nitrate below the root zone. By use of a properly designed sprinkler irrigation system, relatively high water distribution efficiencies, water application efficiency, and water storage efficiency may be obtained. Since high efficiencies are generally more readily obtainable for a sprinkler irrigation system, the nitrate pollution potential may be reduced.

Drip Irrigation

Another irrigation method that has distinctive characteristics of direction and relative extent of water movement is drip irrigation. Figure 4-6 shows a diagram of the direction of water movement under a single emitter. Whether or not the wetted area below the soil surface will coalesce with another wetted area

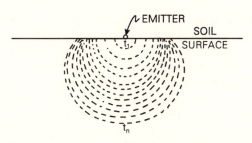

Figure 4-6 A cross-sectional diagram of the direction and relative extent of water movement as a function of time (*t*) for drip irrigation. The actual shape will depend on the soil texture.

depends upon the spacing of the emitters within the field. The general tendency is for water to move radially in all directions from the emitter. However, as the time of water application increases, there is a greater tendency for downward than lateral movement. With frequent irrigations, as is generally the practice under drip irrigation, there is a tendency for nitrates and other solutes to accumulate near the perimeter of the wetted area. This is a function of the water gradient, which is generally away from the emitter.

The importance of recognizing the impact of water movement on placement of fertilizer and nitrogen in relation to the emitter was demonstrated by Miller et al. (1976). When the emitters were placed in the plant row and nitrogen was banded 10 cm deep and 20 cm to the side of the row, the amount of nitrogen obtained by the plant was much less than when either nitrogen was injected through the drip irrigation system or the plants were furrow irrigated. The drip irrigation system tended to move nitrate away from the plant row and the furrow irrigation system moved nitrogen into the plant row where the plant had a greater access to it. There was little difference in nitrogen uptake and use between the furrow irrigated case and when nitrogen application was through the drip system. In both cases, the nitrogen was uniformly distributed throughout the wetted zone.

Water Distribution Under Different Management Techniques

As indicated previously, there are several parameters for judging irrigation efficiency (water application, storage, and distribution efficiency). In Fig. 4-7 are shown profiles of relative water distributions for fields with different surface water management. The diagrams in Fig. 4-7 are applicable to either a flood or a furrow irrigation system when considering an entire management unit. Efficiency values are indicated for three different levels of irrigation application, I_1 being the least amount of water applied and I_3 the greatest. These diagrams also show the influence of tailwater runoff or tailwater reuse systems on the irrigation efficiency parameters. As seen in Fig. 4-7 (C) (D), the same relative distribution of water in the soil profile is obtained whether there is runoff or a tailwater reuse system. The application efficiency is much greater where a tailwater reuse system is employed, but this is the only parameter that is affected by the use of such a recycling system.

The stylized irrigation patterns also indicate the extent of nitrogen movement below the root zone, as well as those locations in the field where nitrate leaching is most likely to occur. One should expect the greatest amount of leaching at the head of the field and the least near the tail-end of the field. Ponding of water on the tail-end of the field will result in greater leaching at this location.

The main purpose of the diagrams is to show that one should not expect the same concentration or mass emission of nitrogen below each portion of the field. This aspect of nitrogen movement is important to recognize when attempting to monitor nitrate leaching below the root zone. If only one location is sampled, it might represent either the maximum or minimum concentration or mass emission. As a result, it is necessary to sample several locations in the field and

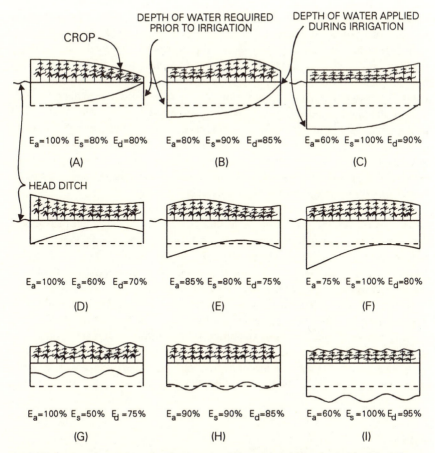

DEPTH OF WATER REQUIRED   DEPTH OF WATER APPLIED
PRIOR TO IRRIGATION   DURING IRRIGATION

CROP

E_a=100% E_s=80% E_d=80% E_a=80% E_s=90% E_d=85% E_a=60% E_s=100% E_d=90%

(A)   (B)   (C)

HEAD DITCH

E_a=100% E_s=60% E_d=70% E_a=85% E_s=80% E_d=75% E_a=75% E_s=100% E_d=80%

(D)   (E)   (F)

E_a=100% E_s=50% E_d=75% E_a=90% E_s=90% E_d=85% E_a=60% E_s=100% E_d=95%

(G)   (H)   (I)

Figure 4-7   Diagram of water distribution in the soil profile of a field with different surface water management systems and efficiencies for application (E_a), storage (E_s), and distribution (E_d) for three levels of irrigation. (Israelsen and Hanson, 1967. Copyright © 1967 by John Wiley & Sons, Inc. Reprinted by permission.)

attempt to integrate the information in order to obtain a properly weighted value for nitrate movement out of the root zone.

Relative distribution of water in field also determines where the nitrogen fertilizer might go if it is applied in the irrigation water. If one expects runoff, a substantial portion of the fertilizer nitrogen applied in the water may run off the end of the field. Assuming a homogeneous mixing of nitrogen in the irrigation water, then the greatest amount of nitrogen will be applied where the greatest amount of water is applied. If the nitrogen added is in the form of nitrate, any water moving below the root zone during the irrigation will leach a portion of the nitrogen that would not be available to the plant. If the nitrogen is in the form of ammonium, then the ammonium becomes adsorbed by soil particles near the surface. The result is a greater concentration of nitrogen near the soil surface where greater total amount of water penetrated the soil. It is evident that the differences that exist in water distribution have an impact on nitrogen placement, utilization, and nitrate leaching potential.

In Figs 4-8, 4-9, and 4-10 are shown the distribution of total nitrate-N and fertilizer nitrate-N as a function of depth and time under different irrigation practices. The fertilizer was applied in bands about 12 cm below the soil surface. Under sprinkler irrigation, nitrate moved in a rather discrete pattern beneath each of the bands. Under furrow irrigation, the nitrate from the bands coalesced as it moved through the soil profile. Under subirrigation, there was no apparent downward movement as the season progressed. This is to be expected since the gradient of water movement is toward the surface. These data point out the relatively high degree of mobility of nitrate in the soil solution and the impact of the irrigation system on nitrogen movement in soils.

Another important concept to be considered is the impact of the irrigation system on the ability to control the leaching fraction and the effect of leaching fraction on the concentration or mass emission of solutes (including nitrate) below the root zone. As shown in Fig. 4-11, as the leaching fraction decreases from 40 to only 10 percent of the applied water, the concentration of salts in the soil-water increases from an electrical conductivity (EC) of about 1.5 dS/m to an EC of nearly 5 dS/m. In other words, more efficient water application substantially increases the concentration of solutes at the bottom of the root zone. Therefore, concentrations of nitrate at the bottom of the root zone, with a 10 percent leaching fraction should be much greater than at the bottom of a root zone with a 40 percent leaching fraction. With higher leaching fractions more nitrogen leaches below the root zone, but the concentrations are also lower.

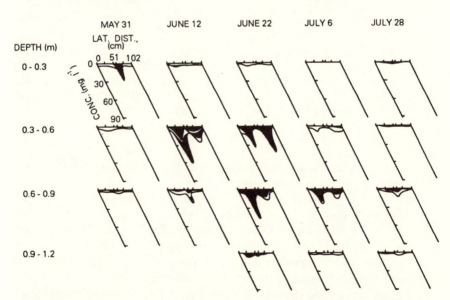

Figure 4-8 Lateral and vertical distribution of total nitrate-N (solid black line) and fertilizer nitrate-N (shaded area) in Miles fine sandy loam (fine-loamy, mixed, thermic udic Paleustalfs) soil sample extracts by date. The plots were sprinkler irrigated and fertilized with ^{15}N-enriched sodium nitrate. (Wendt et al., 1977.)

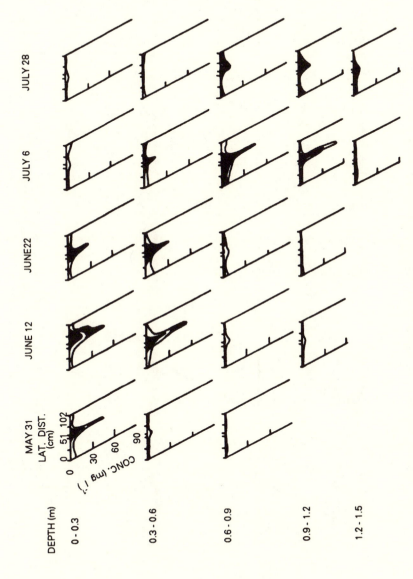

Figure 4-9 Lateral and vertical distribution of total nitrate-N (solid black line) and fertilizer nitrate-N (shaded area) in Miles fine sandy loam (fine-loamy, mixed, thermic udic Paleustalfs) soil sample extracts by date. The plots were furrow irrigated and fertilized with ^{15}N-enriched sodium nitrate. (Wendt et al., 1977.)

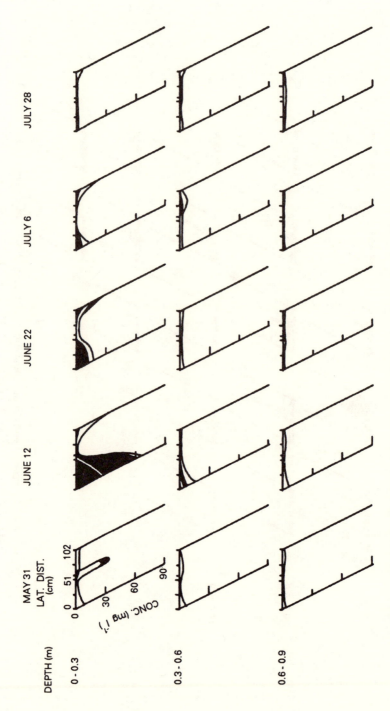

Figure 4-10 Lateral and vertical distribution of total nitrate-N (solid black line) and fertilizer nitrate-N (shaded area) in Miles fine sandy loam (fine-loamy, mixed, thermic udic Paleustalfs) soil sample extracts by date. The plots were subirrigated and fertilized with ^{15}N-enriched sodium nitrate. (Wendt et al., 1977.)

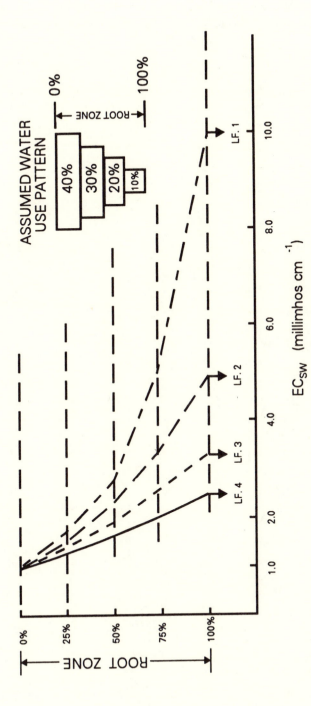

Figure 4-11 Salinity profile expected to develop after long-term use of water of $EC_w = 1.0$ dS/m at various leaching fractions (LF). (Ayers and Westcot, 1985. Copyright © 1985 by the Food and Agriculture Organization of the United Nations. Reprinted by permission.)

Because of the relationships between nitrogen pollution potential and water movement, the irrigation system used in an agricultural production system is a critical factor to consider in evaluating production and pollution potentials.

5

Nitrogen

It is ironic that nitrogen is quite frequently a limiting factor for plant growth while it is one of the most abundant elements in the biosphere. Most nitrogen exists in a form that is not available for plant use. In natural environments and agricultural systems nitrogen is generally deficient for plant growth. However, under some natural conditions nitrogen can accumulate over time to the point where it is adequate for plant growth; but when nitrogen removal increases, such as in agricultural production, then the nitrogen reserve is quickly depleted. In an agricultural system much of the nitrogen is removed in the harvested portion of the crop. Consequently, nitrogen must be added to sustain production.

As shown in Table 5-1, the majority of nitrogen in the biosphere is in the earth's crust and sedimentary deposits (approximately 1.8×10^{16} metric tons). According to Delwiche (1970), this is essentially unavailable for cycling. The next largest quantity of nitrogen exists in the atmosphere. The atmospheric nitrogen pool through biological, industrial, and lightning fixation, is the primary source of nitrogen for biological activity. There must be a continuous cycling of nitrogen between the atmosphere and biological nitrogen pools, otherwise there would be a depleted nitrogen supply for biological activity found on land or in the ocean. The amount of nitrogen that is being fixed yearly by nature or industry is estimated to be in the same order of magnitude as denitrification (Table 5-1). An extremely small portion of the total amount of nitrogen in the biosphere is actually involved in the cycling.

The processes involved in cycling and the nitrogen pools are depicted in Fig. 5-1. The management of processes, transformations, and transfers of nitrogen in soils is important for food production and reduction of pollution potential. Although nitrogen is ubiquitous and in a continual state of change of form, localized concentrations can occur that may be considered undesirable.

By utilizing the biological and industrial fixation of nitrogen, agriculture has been able to meet the food requirements of an increasing population with an improvement of the diet for millions of people currently consuming less than adequate protein. The inevitable processes of nitrogen cycling will make it essential to increase quantities of fixed nitrogen in a form available for plant use.

Table 5-1.   Estimated quantities of nitrogen in the biosphere and involved in cycling.

Nitrogen in biosphere	Amount mt x 10⁹
Earth crust	14,000,000
Sediments	4,000,000
Atmosphere	3,900,000
Land: (Total)	(343)
Organic (abiotic)	175
Inorganic	160
Plants	8
Animals	0.2
Ocean: (Total)	(144)
Organic (abiotic)	45
Inorganic	99
Plants	0.2
Animals	0.2

Nitrogen cycling	Amount mt x 10⁶ per year
Natural fixation:	
Land	99
Ocean	30
Combustion and atmospheric	25
Subtotal	154
Industrial fixation	40
TOTAL FIXATION	194
Denitrification:	
Land	120
Ocean	40
Combustion	50
TOTAL DENITRIFICATION	210

Sources: Delwiche, 1970; Delwiche and Likens, 1977.

BIOLOGICAL TRANSFORMATIONS

Nitrogen Fixation

There are numerous organisms that have the capability of transforming atmospheric dinitrogen into a form that can be utilized for their growth and development. Two main types of biological nitrogen fixation are symbiotic and asymbiotic (Table 5-2). Symbiotic fixation occurs by association of a host plant with a particular species of bacteria. The host plant supplies carbohydrate for the benefit of the bacterium, which fixes nitrogen for the benefit of the host. Asym-

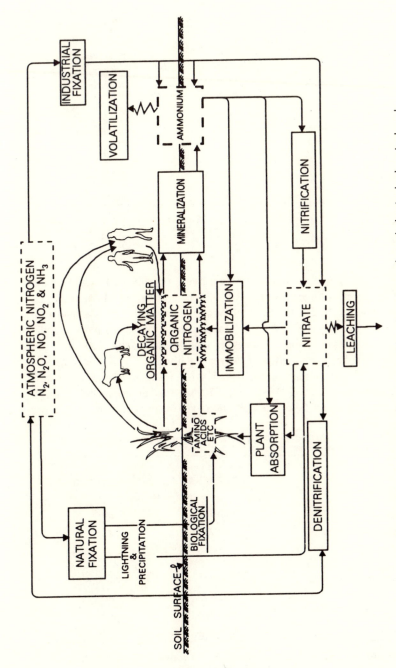

Figure 5-1 Schematic showing various nitrogen pools and physical, chemical and biological processes involved in the cycling of nitrogen through soils. Continuous line boxes represent physical, chemical, and biological processes and/or transformations. Dashed line boxes represent forms and pools of nitrogen.

Table 5-2.   Types of biological nitrogen fixation.

Approximate type	Number of species	Typical organisms	
		Common name	Scientific name
Symbiotic			
Legumes	12,000	Alfalfa	*Medicago sativa*
		Blue lupine	*Lupinus angustifoliüs*
Non-legume	50	Alder	*Alnus* sp
		Lichens	*Collema tenax*
Asymbiotic			
Algae	46	N/A[a]	*Nostoc* sp
		N/A	*Anabaena* sp
Bacteria:	30		
Anaerobic		N/A	*Clostridium* sp
Aerobic		N/A	*Azotobacter* sp

Sources: Nutman, 1965; Hardy and Gibson, 1977; Stewart, 1966.

[a] Common name not applicable.

biotic nitrogen fixation is conducted by free-living organisms. In general, the quantity of symbiotic nitrogen fixation is much greater than asymbiotic nitrogen fixation. As indicated in Table 5-3, symbiotic legume fixation may range from 1 kg/ha to approximately 670 kg/ha per year depending upon the plant and the environmental conditions. Similar ranges in nonleguminous symbiotic nitrogen fixation have also been reported by Hardy and Gibson (1977). Under natural conditions much lower annual rates of fixation for asymbiotic microorganisms have been observed. Steyn and Delwiche (1970) reported approximately 5 kg/ha of N fixed per year under favorable conditions and rates as low as 2 kg/ha per

Table 5-3.   Reported ranges in the quantity of nitrogen fixed by legumes and gains or losses in soil nitrogen.

Legume	Reported range of nitrogen fixed	Reported gain or (loss) in soil N
	----------kg of N/ha/Crop[a]----------	
Alfalfa	50-460	50
White sweetclover	30-203	42
Subterranean clover	45-336	20
Miscellaneous clovers (eg. alsike, ladino, red)	45-673	20-60
Vetch	25-673	40
Dry beans (field)	30-100	(60)
Soybean	1-168	(42)

Sources: Nutman (1965); Hardy and Gibson (1977); Tisdale and Nelson (1966); Cope (1975); Broadbent (1972); and Kroontje and Kehr (1956).

[a] For crops having multiple harvests the amount of nitrogen fixed is for the growing season.

year in more arid sites. In wet mountain meadows, Porter and Grable (1969) found that much of 3.7 to 6.8 kg/ha of N was fixed by algae and anaerobic organisms during a 10-day period. They estimated that approximately 40 kg/ha of N could be fixed during a 100-day growth period where there is a frost-free period of 70 days. The research effort on legumes has been emphasized since these plants are utilized for direct consumption by humans or animals for their protein. Table 5-3 shows some ranges of nitrogen fixed by selected legumes.

Frequently legumes require inoculation by specific organisms before they are capable of fixing nitrogen. In Table 5-4 are listed some legume host groups with their respective symbiotic nitrogen-fixing organism. While it is possible for some of the *Rhizobium* species to cross-inoculate different legume host groups, the *Rhizobium* species indicated for each group is the most effective for the legumes in that group. The effectiveness of these organisms and their response to environmental factors lead to the great range in amounts of fixed nitrogen observed in Table 5-3.

There are also large differences between the total amount of nitrogen fixed during a cropping season and the amount of nitrogen that may be returned to

Table 5-4.   Legume host groups and their associated rhizobium SP inoculant.

Host common name	Scientific name	Rhizobium species
Alfalfa group		R. meliloti
Alfalfa	Medicago sativa	
Tifton Bur Clover	M. rigidula	
White Sweetclover	Melilotus alba	
Clover groups		R. trifolii
Alsike Clover	Trifolium hydridum	
Ladino Clover	T. repens	
Sub Clover	T. subterraneum	
Pea and vetch group		R. leguminosoruses
Garden Pea	Pisum satavium	
Common Vetch	Vicia satavia	
Lentil	Lens culinaris	
Cowpea group		R. spp.
Common Lespedeza	Lespedeza striat	
Peanut	Arachis hypogaea	
Lima Bean	Phasedus lunatus	
Bean group		R. phaseoli
Dry Beans	Phaseolus vulgaris	
Lupini group		R. lupini
Blue Lupine	Lupinus angustifolius	
Soybean group		R. japonicum
Soybean	Glycine max	

Sources: Graham, 1976; Tisdale and Nelson, 1966; and Erdman, 1959.

the soil. The discrepancy occurs as a result of removal of all or a portion of the above-ground crop. For example in Table 5-5, cutting frequency of alfalfa has little effect on the total amount of nitrogen added to the soil. Where similar cutting practices were used for sweet clover and vetch, there was a large reduction in the amount of nitrogen returned to the soil. As shown in Table 5-3, growing beans and soybeans resulted in a net loss of nitrogen from the soil.

The distribution of nitrogen between the root and tops for various other legumes influences the total amount of nitrogen returned to the soil environment. Table 5-6 shows the range in percentage of total nitrogen in tops, roots, and nodules for three forage legumes. While differences exist in the amount of nitrogen returned to the soil by root systems of different crops, the incorporation of the top provides the greatest nitrogen additions.

Table 5-5.   Influence of cutting frequency on the amount of nitrogen added to soil in tops and roots.

Crop/variety	Number of cuttings	Top and root nitrogen added to soil	Yield of tops	Top:root ratio
Alfalfa:		-kg of N/ha-	-mt/ha-	
Arizona common	1	85	2.87	1.56
	2	91	1.46	0.74
	3	71	0.63	0.31
Ranger	1	77	2.71	2.05
	2	78	1.23	0.69
	3	59	0.27	0.14
Sweet clover:				
Madrid	1	203	5.02	1.43
	2	104	0.85	0.25
	3	31	--[a]	--[b]
Hubam	1	58	4.26	3.12
	2	20	0.43	1.84
	3	9	--	--
Vetch:				
Madison	1	108	3.54	10.39
	2	26	0.63	11.43
	3	7	--	--
Hairy	1	89	2.80	6.36
	2	31	0.49	1.66
	3	15	--	--

Source: Abridged from Kroontje and Kehr, 1956. Copyright © 1956 by American Society of Agronomy. Reprinted by permission.

[a] No measurable yield of tops.

[b] Some plant roots still remained, but unable to calculate a top:root ratio.

Table 5-6.   Distribution of nitrogen between tops and roots of root-nodulated forage legumes.

| Legume | Range in percentage of total nitrogen | | |
	Tops	Roots	Nodules
Alfalfa	50-66	29-43	2.5-4.5
Subterranean clover	69-85	11-21	3.6-12.6
Ladino clover	81-84	11-14	3.5-4.4

Source: Stewart, 1966. Copyright © 1966 Athlone Press. Reprinted by permission.

Environmental factors such as temperature, soil pH, aeration, and nitrate concentration in the soil influence the effectiveness of nitrogen fixation. At temperatures suitable for plant growth, nitrogen fixation usually proceeds at an optimum rate. At temperatures above 30°C activity is reduced, and as it exceeds 35°C nodules are generally sloughed. Maximum rates of fixation occur in pH ranges similar to those that are optimum for plants (pH 6 to 8).

Symbiotic nitrogen fixation in legumes has been demonstrated to be inhibited in the presence of oxygen. Yet, low oxygen levels in the soil causes sloughing of the nodules. This is attributed to the influence of low oxygen levels on the host plant, rather than a direct affect on the nitrogen fixation capability of the *Rhizobium* bacteroid. The bacteroid apparently has a mechanism that creates an anoxic condition within the tissue of the host plant, thereby allowing nitrogen fixation to proceed within the roots in an aerobic environment. The extent of anoxia in soils determines which of the asymbiotic organisms may be fixing nitrogen. Under aerated conditions, *Azotobacter* species are capable of fixing nitrogen, and under anoxic conditions *Clostridium* species are capable of fixing nitrogen.

The concentration of nitrate in the soil solution affects symbiotic nitrogen fixation. At concentrations of about 3 ppm nitrate-N in solution culture, Munns (1977) showed that nodulation is delayed in alfalfa. McAuliffe et al. (1958) found approximately 65 and 10 percent of the nitrogen in Ladino clover was biologically fixed at nitrogen rates of 30 and 220 kg/ha of N, respectively. They also found that alfalfa fixed approximately 20 percent of its nitrogen requirement regardless of fertilizer nitrogen applications ranging from approximately 20 to 85 kg/ha.

Immobilization

Immobilization is the process whereby inorganic nitrogen in the soil is incorporated into organic matter and rendered essentially unavailable for utilization by plants. It is a biological process influenced by the carbon–nitrogen ratio of the added organic matter, temperature and pH of the soil, and the chemical form of the available inorganic nitrogen.

This process is carried out by a variety of soil microorganisms capable of using organic carbon as an energy source (Broadbent and Tyler, 1962; Stewart et al., 1963a,b; Broadbent and Nakashima, 1967). It has been shown that a significant portion of immobilized nitrogen remains in the organic fraction for long periods of time. Under conditions of optimum temperature and moisture, Broadbent and Nakashima (1967) found, after 2 years of incubation, that more than half of the added fertilizer was still present in the organic fraction of the soil. An important factor determining whether there is net immobilization or net mineralization is the C : N ratio of the organic material. As previously discussed, high C : N ratios lead to net immobilization of soil nitrogen and low C : N ratios lead to net mineralization. The processes of immobilization and mineralization are occurring simultaneously in soils; consequently one must think in terms of net mineralization or immobilization to understand whether nitrogen will be available or unavailable for plant use.

The amount of nitrogen immobilization occurring is also influenced by inorganic nitrogen sources. A study by Broadbent and Tyler (1965) showed that heterotrophic organisms carrying out the immobilization definitely exhibit a preference for the ammonium source of nitrogen. In Fig. 5-2 are data showing the amount of nitrogen immobilized into the organic form when ammonium and nitrate nitrogen were added along with barley straw. Simpson and Freney (1967) observed the preference for ammonium source during immobilization. In their experiments there was little evidence of remineralization of nitrogen that had been immobilized after 6 weeks of incubation. The influence of soil salinity on the rate of immobilization was demonstrated by Westerman and Tucker (1974). Their data indicate that a 100-fold increase in salinity decreased ammonium immobilization from 95 to 63 percent of the added ammonium nitrogen.

Mineralization

Mineralization is the transformation of organic nitrogen into an inorganic form, (ammonium-N + nitrate-N). This process is common to all types of hetero-

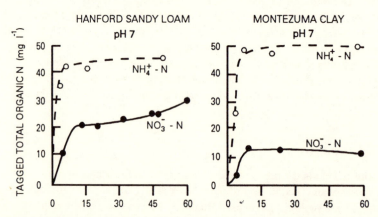

Figure 5-2 Influence of form of inorganic nitrogen on net immobilization in Hanford sandy loam (coarse-loamy, mixed, nonacid, thermic type xerorthents) and Montezume clay (ashy, mesic haploxerollic Durorthids) soils in California. (Broadbent and Tyler, 1965. Copyright © 1965 by M. Nijhoff. Reprinted by permission.)

trophic microorganisms. Mineralization is also affected by the C : N ratio of the organic material. Rubins and Bear (1942) found that C : N ratio alone was not a reliable means of determining whether net mineralization will occur. The C : N ratio coupled with the nitrogen content was more indicative of net mineralization. This accounts for the rather wide range of C : N ratios at which net mineralization occurred as reported by numerous investigators. As indicated in Tisdale et al. (1985), the C : N ratio may change from approximately 60 : 1 to less than 20 : 1 when decomposition of organic material is allowed to proceed for 4 to 6 weeks. Allison (1961) and Rubins and Bear (1942), among other investigators, have indicated that the amount of carbon in a readily available form for microbial use determines how rapidly the C : N ratio may change from one of net immobilization to net mineralization. These observations are confirmed by the work of Sain and Broadbent (1977), showing no effect of nitrogen addition on total decomposition after a period of 120 days. Powers (1968) has also shown that the amount of net mineralization is proportional to the amount of material with low or high nitrogen content (0.8 and 1.4 percent, respectively). He also shows that the decomposition pattern of the plant residue is determined primarily by the composition of the residue rather than soil properties such as organic matter content, pH, and fertility levels.

Numerous investigators have found that nitrogen is ordinarily released from the stable organic matter fraction of soils at the rate of 1 to 3 percent during a growing season (Allison, 1956). Stanford and Smith (1972) reported 5 to 41 percent mineralization for 39 soils from several locations, and Stanford et al. (1974) indicated a range of 14 to 19 percent mineralization of nitrogen in several Idaho soils. In the surface 15 cm, this rate of mineralization can contribute approximately 200 kg/ha of N for a sugar beet crop (Carter et al., 1974). Assuming a 6 percent nitrogen content in the organic matter fraction of the soil, this would correspond to approximately 0.6 to 0.7 percent organic matter in the soil, which is not an unrealistic value for arid soils in the western United States.

The apparent discrepancy between the relatively high rate of nitrogen mineralization as found by Stanford et al. (1974) and Carter et al. (1974) in Idaho and the relatively low rate of mineralization as indicated by other investigators may be partially attributed to the cropping sequence generally followed in Idaho. Dry beans and alfalfa are legumes that figure prominently in the crop rotation system employed by the farmers of the area.

Changes in the cropping sequence or cultural practices resulting in a shift in the amount of crop residue returned to the soil may change the amount of mineralizable nitrogen. When applications of the crop residue are made, the cumulative nitrogen contribution from the crop residue approaches the total amount of nitrogen being incorporated in the crop residue as the process nears steady-state conditions. According to Pratt et al. (1973), the rate of mineralization of nitrogen from various materials approaches steady-state conditions in about 20 years. This effect is somewhat corroborated by Mayurak and Conard (1966) who reported rapid loss of organic matter from soils during the years immediately following the breaking of native sod. The rate of loss became progressively less with time.

In rather large areas of the irrigated west, the fall and winter temperatures are warm enough that nitrogen mineralization will still occur. Hipp and Gerard (1971) found that after a winter fallow there was a substantial increase in nitrate-N content in a fine sandy loam soil as opposed to where a winter crop had been grown. They found 80 kg/ha more nitrate-N in the 120-cm deep soil profile of the fallow area than in the cropped area. They concluded the increase in nitrate-N in the soil profile was due to fall and winter mineralization since the soil profile was essentially depleted of nitrate-N at the fall harvest of the previous crop.

Nitrification

Another of the key transformations of nitrogen in the soils is nitrification. This process is a biological oxidation starting with ammonium-N through a series of intermediates to the nitrate form. The most widely recognized intermediate is nitrate-N which can accumulate in the soil environment depending upon certain conditions. Microorganisms from the genera *Nitrosomonus* and *Nitrococcus* are principally associated with the transformation of ammonium to nitrite. *Nitrobacter* organisms are those associated with the transformation of nitrite to nitrate. The overall reaction is as follows:

$$2NH_4^+ + 3O_2 \rightarrow 2NO_2^- + 2H_2O + 4H^+$$

$$2NO_2^- + O_2 \rightarrow 2NO_3^-$$

Oxygen is required for the reaction to proceed, with approximately 20 percent oxygen in the soil air required for maximum nitrification (Amer, 1949). However, only a slight decrease in the rate of nitrification occurs as the oxygen content decreases to approximately 11 percent. Similar results were observed by Patrick and Gotoh (1974) while investigating the influence of atmospheric oxygen concentration on nitrification in flooded soils. Patrick and Sturgis (1955) have shown that ambient atmospheric oxygen concentration yielded dissolved oxygen levels of 6 to 8 ppm in water above the surface of a flooded soil. As pointed out by Jenkins (1969) in his study of nitrification in effluents, dissolved oxygen concentrations approaching 0.5 ppm appear to be the lower limit for nitrification.

The release of hydrogen ions to the soil environment from the nitrification reactions may contribute to soil acidification. Broadbent et al. (1958b) followed the pH changes in six soils as a function of different nitrogen sources. At nitrogen application rates comparable to what is commonly practiced in fertilization of crops, the alkaline and well-buffered soils (either alkaline or slightly acid) showed relatively little change in pH over the 6-week study period.

The rate of nitrification is also markedly affected by the soil temperature (Tyler et al., 1959; Parker and Larson, 1962; Sabey et al., 1969). Figure 5-3 adapted from Parker and Larson (1962) shows the amount of nitrate-N produced during a 35-day incubation period for different soil temperatures when 150 ppm of ammonium-N had been added to a silt loam soil. Temperature affected the rate of nitrification and the amount of time required to reach maximum nitrifi-

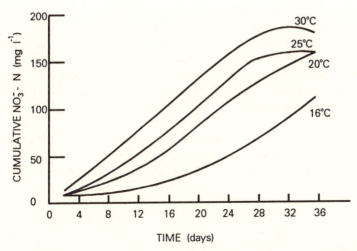

Figure 5-3 Influence of temperature on rate of nitrification in a silt loam soil. (Parker and Larson, 1962. Copyright © 1962 by the Soil Science Society of America. Reprinted by permission.)

cation. Tyler et al. (1959) found that four soils receiving 224 kg/ha ammonium-N as aqua-ammonium or ammonium sulfate had similar rates of nitrification at 24°C. At lower temperatures, the soil differences affected the rate of nitrification to a greater extent. This is probably due to a resting population of nitrifiers, as pointed out by McLaren (1969). Nitrification was observed at temperatures as low as 3°C, although at a very reduced rate when compared to either 7°C or 24°C (Tyler et al., 1959). At normal nitrogen rates with uniform application, they found that 4 to 9 kg/ha per day of ammonium-N was nitrified at 7°C and from 25 to 30 kg/ha per dat at 24°C.

Extremes of high or low pH will also inhibit soil nitrification. After a band application of anhydrous ammonia soil pH changed from 6.7 to 8.3 and 9.1 for air dry and moist soil, respectively, in 36 hours (Papendick and Parr, 1966). Tisdale and Nelson (1966) and Broadbent et al. (1958b) indicate that nitrification can proceed even at pH values above 9 (up to about pH 10, with optimum pH for nitrification at about 8.5). They also indicate that below pH 5.5 nitrification proceeds at a much reduced rate with some nitrification occurring at pH values as low as 3.8. This indicates that nitrification can occur over a wide range of pH conditions and one would not expect any great limitations on nitrification in normal agricultural soils.

Several investigators have examined the possibility of reducing the rate of nitrification by addition of a chemical to the soil along with the ammonium fertilizers. Pioneering investigations resulted in the discovery of a material known as 2-chloro-6-(trichloromethyl)pyridine, which has been found to inhibit the *Nitrosomonas* organisms that transform ammonium to nitrite. Goring (1962) reported the effects of this material on the rate of nitrification of various ammonium fertilizers in 21 different soils from throughout the United States representing numerous soil textures. This material was shown to be effective in reducing the rate of nitrification. Turner et al. (1962) found that there was a general trend toward higher levels of ammonium recovery as the concentration of this inhibitor increased. Other investigators (Sabey, 1968; Patrick et al., 1968; Moore and Soltanpour, 1974; Papendick et al., 1971) have investigated potassium azides and 2-amino-4-chloro-6-methylpyridine as other compounds that may have economic potential as nitrification suppressants. The general conclusion of these investigators is that 2-chloro-6-(trichloromethyl)pyridine was the most effective.

The nitrification process may also be inhibited by various pesticides. Chandra and Bollen (1961) found that certain fungicides completely repressed nitrification for 30 days in the laboratory. In laboratory studies with cell suspensions, Winely and San Clemente (1969) found that *Nitrobacter agillis* was substantially inhibited in the presence of an insecticide at relatively high concentrations. The fungicides used by Chandra and Bollen (1961) were applied at rates normally used for soil fumigation. Consequently, in fumigated soils nitrification may be suppressed.

Nitrification is a relatively rapid process that occurs over a wide range of environmental conditions resulting in the ultimate formation of nitrate in the soil. This frequently leads to the conclusion that plants only utilize the nitrate form of nitrogen, since that is the available form usually detected in soils. However, it has been amply demonstrated (see Nitrogen Assimilation in Chapter 3) that plants have the capability of utilizing the ammonium and nitrate forms of nitrogen. Because of the nitrification process, nitrate is the predominant form of available nitrogen for plant use.

Denitrification

Denitrification is one of the major mechanisms of nitrogen loss from the biosphere. The principal mechanism is a biological process involving the reduction of nitrate to gaseous end-products. Many species of microorganisms are capable of carrying out the nitrate reduction process. Jordan et al. (1967) discussed the results of numerous investigators that showed denitrifying microorganisms to be ubiquitous in soil and water environments. They were able to identify 22 species of microorganisms capable of reducing nitrate to nitrite. Some of these organisms were incapable of further reducing nitrite to gaseous end-products. The ubiquitous nature of microorganisms that have the capability of further reducing nitrite makes it unlikely that nitrite would accumulate in soils during the denitrification process except under unusual conditions.

Although the denitrification process is not very specific with regard to the organisms involved, it takes place only under very specific conditions. The conditions for denitrification must occur simultaneously in the soil environment, otherwise the process cannot proceed. The essential conditions are: (1) A drastic lowering of the oxygen level in the soil gases and soil solution. This is generally related to a lowering of the oxidation–reduction potential of the soil environment to below +225 millivolts at pH 7. (2) An adequate supply of carbonaceous energy source for the heterotrophic soil organisms. The organisms that carry out the denitrification are facultative aerobes. They utilize oxygen in nitrate as an electron acceptor during decomposition of organic matter under anaerobic conditions. (3) The organism must be present. This third condition is met for most if not all soils. The nitrate concentration does not seem to materially affect the rate at which denitrification occurs, as shown by Cooper and Smith (1963), Patrick (1960), and McLaren (1976).

The sequence of denitrification is given as follows:

$$\text{Nitrate (NO}_3^-\text{)} \rightarrow \text{nitrite (NO}_2^-\text{)} \rightarrow \text{nitrous oxide (N}_2\text{O)}$$
$$\rightarrow \text{nitric oxide (NO)} \rightarrow \text{dinitrogen (N}_2\text{)}$$

The last three compounds are gaseous, with their relative ratios of production depending upon a number of environmental factors. The two principal nitrogen gases produced are N_2O and N_2. Numerous investigators have shown that the $N_2 : N_2O$ ratio varies as a function of time, nitrate concentration, energy supply, degree of anoxia, and the type of organisms present. The relative impact of each of these factors, except nitrate concentration, on the $N_2 : N_2O$ ratio and the total amount of denitrification is reported by Rolston (1977). He shows the influence of the presence of carbonaceous material and the degree of soil saturation on the $N_2 : N_2O$ ratio and total amount of denitrification (Table 5-7). The most rapid denitrification took place in the first 6 to 24 h after application of the nitrate. No denitrification products were detectable above background concentrations after 15 to 25 days. In the wettest plots that received manure, the greatest concentrations of N_2 and N_2O were measured at the 2-cm sampling depth below the surface. He concluded there must be pockets of anoxic conditions within the soil profile even near the surface. Such conditions could be created by high

Table 5-7. Influence of organic carbon source and soil-water regime on denitrification measured in the fields.

Treatments			Total denitrification[c]	
Carbon source[a]	Soil-water[b] regime	$N_2:N_2O$ ratio	Amount	Percent of fertilizer
		kg N/ha	kg/ha	%
Manure	h = -15 cm	20 : 1	208	69
Manure	h = -70 cm	7.78 : 1	47	16
Cropped	h = -15 cm	6.98 : 1	34	11
Cropped	h = -70 cm	3.89 : 1	9	3
Uncropped	h = -15 cm	2.71 : 1	8	3
Uncropped	h = -70 cm	6 : 1	4	1

Source: Adapted from Rolston et al., 1977.

[a] Manure at the rate of 50 mt/ha, perennial rye grass and uncropped Yolo loam soil.

[b] Soil-water pressure heads (h), -15 cm and -70 cm indicates two soil-water levels near saturation: e.g., field capacity is approximately -330 cm.

[c] Isotopically labeled nitrogen fertilizer (KNO_3) was used and the soil temperature averaged 23°C.

microbial activity resulting from the input of carbonaceous material and a slow rate of oxygen diffusion to such sites. Later studies by Parkin (1987) and Staley et al. (1990) have demonstrated the presence of "hot spots" of microbial activity in the soil when a readily available carbon source is added. Rolston (1977) also found greater amounts of N$_2$O being produced as a denitrification end-product in the sampling periods immediately following application of a nitrate fertilizer. Perhaps this can be attributed to the presence of nitrate in concentrations high enough to inhibit conversion of N$_2$O to N$_2$.

The relationships between oxygen concentration in soils and the redox potential have permitted the measurement of redox potential to be used as a method of determining denitrification potential of the soil environment. Engler et al. (1976) showed that the redox potential dropped markedly at the point of oxygen disappearance from a soil suspension. Meek et al. (1969) measured the influence of soil-water saturation of soil on denitrification and related it in turn to redox potential. Figure 5-4 shows the relationship between oxygen consumption, redox potential, and the rate of denitrification. Similar relationships between the oxygen concentration an the redox potential have been shown by Reddy and Patrick (1976). When they investigated the effect of the length of the aerobic and anoxic cycles on the amount of denitrification, they found mineralization occurred during the aerobic cycle and denitrification occurred during the anoxic cycle. A considerable amount of native and organic nitrogen was lost during the 128-day incubation period.

An interesting aspect of the denitrification studies in recent years has been the revelation that relatively large amounts of denitrification are occurring in presumably aerobic soils. Field data collected by Broadbent and Carlton (1976) show that nitrogen losses have amounted to from 16 to 25 percent of the total amount of fertilizer nitrogen added each year over a 3-year period. Since sam-

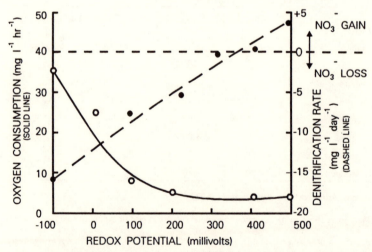

Figure 5-4   Relationship between oxygen consumption, redox potential and rate of denitrification in suspensions of a Crowley Silt Loam (fine, montmorillonitic, thermic typic Albaqualfs) at pH 5.0. (Patrick, 1960.)

pling was done to a depth of 450 cm and the concentration of isotopically labeled N at the lower depth was low, it was assumed the major portion of loss was due to denitrification. Since denitrification is taking place, it must be attributed to the fact that soil microbial activity and slow oxygen diffusion creates microsites of anoxic conditions where denitrification can occur.

Potential denitrification activity was concentrated in the 0–3.8 cm surface layer of soil and decreased to barely detectable levels in the 15–30 cm layer. Significant differences were found as a function of tillage (Staley et al., 1990). The no-till treatment increased denitrification activity when compared with conventional tillage, which they report has been demonstrated by several other investigators. They speculated that a contributing factor causing the difference is the presence of a greater amount of oxidizable carbon in the surface soils under no-till conditions. Parkin (1987) found that "hot-spots" of high denitrification activity were associated with the presence of particulate organic carbon in the soil. These hot-spots had denitrification activity several orders of magnitude greater than the inorganic soil material. The presence of readily decomposable organic carbon maintained a high rate of oxygen consumption, thereby creating an anoxic condition in microsites within the soil profile.

The importance of soil texture on denitrification has been evaluated by Lund and Elliot (1976) and Pratt et al. (1976). They showed that increasing clay content in the 10 to 40 cm depth of the soil profile (control section) had an influence on the saturated hydraulic conductivity and a concomitant effect on the nitrate-N to chloride ratio. Where denitrification is occurring, it is presumed that the nitrate-N to chloride ratio will be less than unity. While this hypothesis continues to be evaluated, it does appear that the potential for denitrification can be correlated with the soil particle size in the control section.

PHYSICAL LOSSES OF NITROGEN

Volatilization

Losses of nitrogen by denitrification are sometimes included with volatilization losses since a gas is produced that subsequently volatilizes to the atmosphere. However, for purposes of this discussion, volatilization will be considered to include only the physical loss of ammonia to the atmosphere. These losses occur when organic material or ammonium fertilizers are placed on or near the soil surface. A source of the commercial fertilizer, urea, is sometimes considered as an ammonium source since it is rapidly hydrolyzed by an enzyme in soil to form ammonium. Ammonium undergoes chemical reactions in the soils, lending to the formation of free ammonia and subsequent volatilization to the atmosphere.

The mechanism whereby ammonia volatilizes from microenvironments due to application of ammonia or ammonia-forming compounds to soils was elucidated by Fenn and Kissel (1973). They showed the total amount of ammonia volatilized was related to the solubility product of calcium with the anion accompanying the ammonium in the fertilizer. Precipitation of calcium compounds with low solubility causes the formation of ammonium hydroxide (NH_4OH), resulting in an increase in soil pH and subsequent dissociation of NH_4OH to yield $NH_3 + H_2O$. The ammonia is then subject to volatilization loss. Previous studies by Martin and Chapman (1951) and Terman and Hunt (1964) alluded to the influence of calcium on the magnitude of ammonia volatilization loss but did not further elucidate the mechanism. In Fig. 5-5 is shown the relative ammonia volatilization loss from ammonium compounds with different accompanying anions. Although some of these compounds are not used as fertilizer materials, they were used in these research investigations to demonstrate the mechanism governing the ammonia volatilization losses. Rapid soil pH shifts were observed where relatively insoluble compounds were formed. The influence of the pH shift after application of these materials on the rate of ammonium loss per hour is shown in Fig. 5-6.

For Dixon silt loam soil at pH 6.5, Ernst and Massey (1960) found a rather low rate of ammonia loss initially with subsequently a more rapid loss during the next 2 to 4 days over a wide temperature range. This is shown in Fig. 5-7, where urea was used as the nitrogen source. It is presumed the delay in volatilization loss is due to a combination of factors including the soil pH below 7 and a delayed hydrolysis of urea. The fact that much lower loss from the surface was recorded than has been reported by Volk (1961), Martin and Chapman (1951), and Terman and Hunt (1964) was probably a function of lower initial soil pH. Ernst and Massey (1960) also showed ammonium losses of 8 and 50 percent from soils pH of 5 to 7.5 after 10 days. Similar results have been shown by Clay et al. (1990) on a soil at pH 6.5 where urea was applied in solution to the surface of the soil. Ammonium volatilization occurred in a period from 2 to 3.5 days after application. These data serve to illustrate the marked effect of initial soil pH on volatilization losses.

The buffering capacity of the soil has also been shown by Martin and Chapman (1951) to determine the extent of nitrogen losses from ammonium materials added to the soil. They concluded from their studies that poorly buffered acid soils allowed the pH to shift sufficiently to permit volatilization of ammonia. Volk (1961) showed an average volatilization loss of 29 percent in 2 weeks when urea was applied to a Leon fine sand at pH 5.8.

In instances where calcium carbonate has been added to acid soils to increase pH, the higher pH resulting from liming causes an increased loss due to volatilization (Volk, 1961; Martin and Chapman, 1951). Fenn (1975) also indicated that mixing monoammonium phosphate with ammonium fluoride, ammonium sulphate, or ammonium carbonate decreased the magnitude of volatilization losses. Studies conducted by Bremner and Douglas (1971) showed a marked reduction in ammonia loss from urea when phosphoric acid was mixed with urea. Kresge and Satchell (1959) found a reduction of ammonia loss by 40 to 50 percent when a mixture of urea with about 30 percent ammonium nitrate was used.

Other factors such as soil-water content, amount of irrigation water applied, temperature, cation exchange capacity, and rate of nitrogen application also influence the amount of loss that will occur. The influence of temperature and rate of nitrogen application can be seen in Fig. 5-8. For ammonium nitrate, there

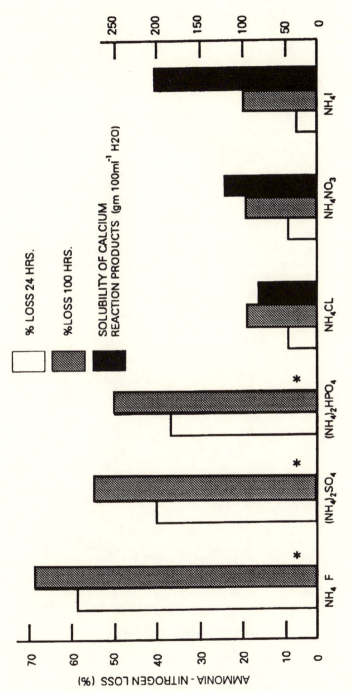

Figure 5-5 The influence of the accompanying anion and its calcium reaction product solubility on the amount of ammonia volatilization loss in 24 and 100 hours after application of the ammonium compound at a rate equivalent to 550 kg/ha of ammonia-nitrogen. Asterisk indicates solubility much less than 1 g/100 ml water. (Adapted from Fenn and Kissel, 1973. Copyright © 1973 by the Soil Science Society of America. Reprinted by permission.)

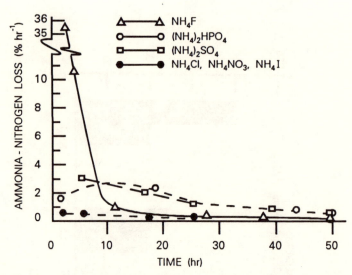

Figure 5-6 The rate of ammonium volatilization loss from Houston Black Clay (fine, montmorillonitic, thermic udic Pellusterts) soil (pH 7.8) as influenced by the anion of the ammonium salt. (Fenn and Kissel, 1973. Copyright © 1973 by the Soil Science Society of America. Reprinted by permission.)

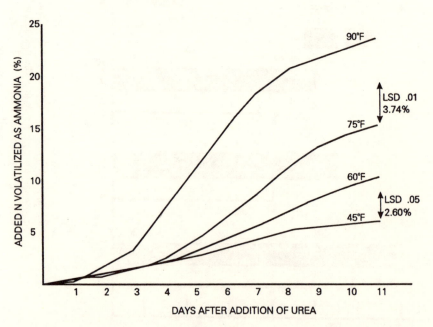

Figure 5-7 Cumulative loss of added N from urea as affected by temperature. Urea-N was top-dressed at a rate of 112 kg/ha on a Dickson silt loam soil (fine-silty, siliceous, thermic glossic Fragiudults) with a pH of 6.5 and aerated at four constant temperatures. (Ernst and Massey, 1960. Copyright © 1960 by the Soil Science Society of America. Reprinted by permission.)

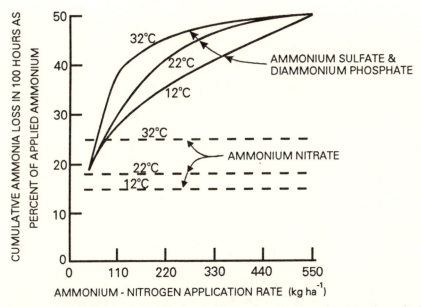

Figure 5-8 Influence of temperature, nitrogen source, and rate of applied ammonium-nitrogen on ammonia volatilization loss from the surface of a Houston Black Clay (fine, montmorillonitic, thermic udic Pellusterts) at pH 7.8. (Fenn and Kissel, 1974. Copyright © 1974 by the Soil Science Society of America. Reprinted by permission.)

is little or no effect of rate of application on the amount of ammonium loss. Higher temperature does increase the amount of loss. For ammonium sulfate and diammonium phosphate materials, temperature and rate affect the magnitude of loss.

Other factors such as cation exchange capacity (CEC) and amount of irrigation have also been examined by Fenn and Kissel (1976) and Fenn and Escarzaga (1976, 1977) for their impact on ammonia volatilization loss. The impact was greatest when ammonium was placed 7.5 cm below the soil surface. Applications of ammonium with irrigation to dry soils resulted in losses on the order of 30 percent of applied nitrogen. When applied to moist soils, the loss increased to nearly 80 percent. They also have shown that application of ammonium nitrate or ammonium sulfate in the irrigation water will result in approximately a 5 percent increase in the magnitude of ammonia volatilization loss in comparison to the dry material surface-applied and then irrigated. Approximately 5 cm water had to be applied to achieve the minimum ammonium loss.

In a study by Henderson et al. (1955) a 50 to 60 percent loss of ammonia occurs with sprinkler irrigation containing about 70 ppm ammonium-N. Even at 9 ppm there was about 35 percent ammonia loss. The influence of solution pH on the magnitude of the volatilization loss is shown in Fig. 5-9. Myamoto et al. (1975) found there was little change in the magnitude of the ammonia loss as concentrations of ammonia increased from 50 to 200 mg/l. The magnitude of loss remained about 50 percent. This agrees relatively well with the results

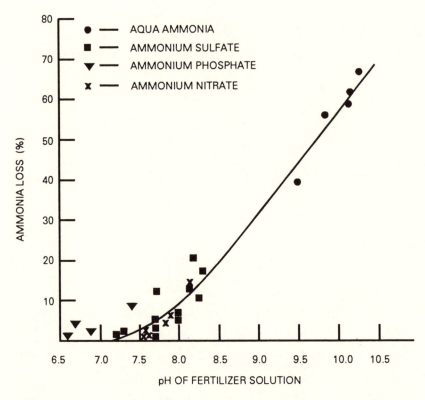

Figure 5-9 Losses of ammonia from aqua-ammonia and ammonium salts in relation to the pH of the fertilizer solution. (Henderson et al., 1955. Copyright © 1955 by the American Society of Agricultural Engineers. Reprinted by permission.)

of Henderson et al. (1955). Additions of sulfuric acid decreased the magnitude of the ammonia loss from 50 to 15 percent (Miyamoto et al., 1975).

The two factors having major impact on the magnitude of volatilization loss from surface-applied ammonia or ammonia-forming compounds are nitrogen source and soil pH. The loss can essentially be eliminated by placement of the fertilizer material below the soil surface. The greater the depth of incorporation, the greater the reduction in volatilization loss. If surface application of ammoniacal fertilizer is a cultural practice that cannot be avoided, then such factors as soil moisture, temperature, CEC, and length of time before either mechanical or irrigation incorporation can be accomplished after the application are all factors that influence the relative magnitude of the loss for a given ammoniacal source and soil pH. In general, cool temperatures, dry soil, CEC in excess of 20 milliequivalents per 100 grams of soil, and application of water in quantities ranging from 2 to 5 cm reduce the magnitude of volatilization loss. These practices serve only to reduce the loss and not to eliminate it. The fact remains that the only way to eliminate loss is to incorporate the ammonium source below the 10-cm soil depth. Several of these factors and their effect on the magnitude of ammonia volatilization losses have been elucidated by Rolston (1978).

Leaching

Leaching of nitrate from the crop root zone is one of the principal concerns of agriculturalists and environmentalists alike. Once beyond the root zone it is no longer available to plants or undergoes transformations, which can reduce the quantities that may reach surface and groundwater supplies. Leaching is the physical process of the downward movement of dissolved constituents in soil solution. As the soil solution is displaced through the soil profile by rainfall or irrigation in excess of the water-holding capacity of the soil, the dissolved ions move with the wetting front. It is the chemical nature of the nitrogenous compounds, nitrogen transformations, and reactions with soil that regulates nitrogen mobility and the quantity that can be leached.

In any soil–plant–water system, nitrate-N movement is the ultimate result of nitrogen transformation in soils and presence of water; thus the potential for movement below the plant root zone always exists. Such movement has been demonstrated many times over the past 100 years. A rather comprehensive report on the amount of nitrates in the drainage waters for a 35-year period at the Rothamsted Experiment Station was published by Miller (1906). Studies by Stout and Burau (1967) and Rible et al. (1976) show variations in nitrate-N concentrations in soils and soil solutions with deep profile sampling under a variety of soil–plant–water conditions. The amount of nitrate-N found varied considerably, but in no case was there an absence.

Regardless of whether the nitrogen source is inorganic or organic, nitrates are found in the soil profile. For example, Bielby et al. (1973) and Adriano et al. (1971) have shown nitrate movement below the root zone as a function of the addition of poultry and dairy manure. The potential for nitrate leaching below the root zone is often greater when relatively unavailable organic nitrogen sources are used for fertilization. Since nitrate formation as a result of mineralization continues even after the crop is removed, nitrate will accumulate. This creates a condition where there is a potentially greater amount of residual nitrate available for leaching. This is similar to the situation pointed out by Viets and Aldrich (1973) where delayed release of nitrogen from slow-release fertilizers could actually increase the potential for a nitrate leaching.

One of the problems associated with measuring the amount of nitrate that may be leaching from a soil profile is the extreme variability in nitrate concentration in the soil and soil solution (Rible et al., 1976; Ruess et al., 1977). These results are similar to those obtained by Nielsen et al. (1974), who reported a method of dealing with the spatial variability and along with Warrick et al. (1977) have been developing mathematical models for evaluating the variability and predicting soil nitrogen concentrations and movement.

The amount of irrigation water applied has a major impact on the quantity of residual inorganic nitrogen in the soil. It also influences the amount of nitrate that may be leached below the root zone. The data reported in Table 5-8 indicate that nitrogen, at rates greater than that required to produce maximum yield,

Table 5-8. Inorganic N in soil after harvest in 3 successive years, 0-300 cm, Davis site. Values are means of 8 soil cores ±95% confidence interval.

Irrigation	Fertilizer rate, kg/ha							
	0		90		180		360	
	Mean kg/ha	C.V., %	Mean kg/ha	C.V., %	Mean kg/ha	C.V., %	Mean kg/ha	C.V., %
					1974			
1/3 ET*	122 ± 28	27.4	132 ± 6.3	5.7	198 ± 39	23.6	364 ± 134	44.0
1 ET	123 ± 45	43.8	104 ± 36	41.0	153 ± 15	11.7	331 ± 62	22.4
5/3 ET	121 ± 11	10.9	119 ± 9.1	9.1	153 ± 8.7	6.8	342 ± 68	23.8
					1975			
1/3 ET	110 ± 4.8	5.2	133 ± 15	13.5	184 ± 46	29.9	423 ± 87	24.6
1 ET	120 ± 11	11.0	138 ± 5.5	4.8	139 ± 13	11.2	409 ± 101	29.5
5/3 ET	117 ± 6.4	6.5	137 ± 8.6	7.5	134 ± 6.3	5.6	295 ± 100	40.5
					1976			
1/3 ET	120 ± 9.2	9.2	154 ± 38	29.7	174 ± 31	21.4	605 ± 149	29.5
1 ET	109 ± 3.8	4.2	116 ± 11	11.6	107 ± 6.7	7.5	394 ± 161	48.9
5/3 ET	116 ± 2.8	2.9	116 ± 4.5	4.6	110 ± 7.8	8.5	263 ± 110	50.0

Source: Broadbent and Carlton, 1979.

* ET = Evapotranspiration. Irrigation is supplied to meet 1/3, 1, or 5/3 of evapotranspiration loss.

accumulated in the soil profile for the two drier irrigation treatments. In the high irrigation treatment nitrate accumulation in the soil profile was less at all nitrogen levels except for the 360 kg/ha rate. Since the same rates of nitrogen were applied for all irrigation treatments, it is apparent that nitrate was being leached in large quantities where nitrogen and water were being applied in excess of the amounts required for maximum yield.

In a nitrogen balance study in Nebraska using isotopic nitrogen, Olson (1976) reported that the amount of residual nitrogen in the upper profile was greater with frequent light irrigations. The study was conducted on sandy soils for a 3-year period. In another study, Onken et al. (1977) were able to account for approximately 93, 86, and 51 percent of the fertilizer nitrogen applied to a loamy fine sand soil under sprinkler, furrow, and subirrigation treatments, respectively. The greater loss occurring under the subirrigation system was attributed to denitrification. In terms of plant uptake, 62, 55, and 49 percent of the fertilizer was used by the corn under sprinkler, furrow, and subirrigation systems, respectively. Their experience was similar to Broadbent's (1976) in that more nitrate was leached from the soil profile when irrigation water quantity exceeded evapotranspiration (ET). At irrigation amounts equivalent to ET little or no nitrate was leached. They also found different patterns of nitrate movement for the different irrigation systems. In the furrow-irrigated plots, nitrate concentrations in soil samples below the root zone were higher where fertilizer was banded below the water level in the furrow. They also found that nitrate moved in a discrete vertical path under banded nitrogen during sprinkler irrigation. In the furrow irrigation system, nitrogen from the fertilizer bands moved toward the center of the bed, merged, and moved vertically downward. In the subirrigation system, the fertilizer bands moved up and away from the point of fertilizer application. These data demonstrate that nitrate movement through soils is closely related to nitrogen placement, quantity and frequency of irrigation, and the method of irrigation.

The influence of method of irrigation and placement of nitrogen fertilizer on movement and concentration of nitrate in soils has also been studied by Nielsen and Banks (1960). They showed no accumulation of nitrates in the surface 0 to 15 cm, the middle of the row, or in the bottom of the furrow under sprinkler irrigation. With furrow irrigation there was a considerable reduction of nitrate in the bottom of the furrow and a large increase in nitrate in the 0 to 15-cm depth in the middle of the row. Again, placement and method of irrigation largely regulated the movement and distribution of nitrates in the soil profile.

Leaching of other inorganic sources of nitrogen is of less importance; however, their movement can be important in terms of placement and subsequent nitrification and leaching. For purposes of this discussion, urea will be considered as an inorganic source of nitrogen even though technically it is an organic compound. As indicated previously, urea hydrolysis occurs relatively rapidly in soils. It does not occur instantaneously and, since urea exists initially as a neutral compound, urea undergoes little reaction with soil colloids, allowing it to move more freely in soils. Urea movement through soils has been shown to be dependent upon the depth of water movement, the rapidity of hydrolysis to ammonia, and the extent of its retention by soils. Broadbent et al. (1958a) found that urea lags somewhat behind the wetting front in a typical Gaussian distribution. Consequently, they concluded that there is little danger of urea leaching below the root zone with an irrigation except possibly in coarse textured soils.

The feature of initial urea movement with water through the soil can be used to advantage in order to achieve proper placement of surface-applied urea nitrogen. Surface application and subsequent irrigation or precipitation can move urea sufficiently into the soil profile to prevent volatilization losses. In addition, the location of the urea nitrogen in the root zone means that subsequent ammonium and nitrate should be immediately available for plant use.

The adsorption of ammonium by soils prevents its movement to any appreciable extent through the soil profile. In the five soils ranging from clay to sandy loam texture studied by Broadbent et al. (1958a), over 90 percent of the ammonium nitrogen was found in the first 4 cm of the soil depth. Only in the sandy loam was any noticeable increase in ammonium concentration evident in the next 4-cm increment. In a more recent study, Fenn and Escarzaga (1977) examined the effects of initial water content of a clay loam soil and the amount of applied water on the depth of ammonium movement in the soil columns. They found a slightly greater depth of movement when the soil was initially saturated, and a slight increase in the depth of movement with increasing rate of water application. They found the majority of the ammonium was retained in the top 5 cm of the soil profile.

Surface Runoff

Of all the loss mechanisms for nitrogen, the amount lost in surface runoff is much lower than for any of the others in an irrigated agricultural system. Unfortunately, most of the investigations in irrigated agriculture combine subsurface return flows with surface runoff. In these cases it is impossible to evaluate the nitrogen losses occurring strictly from surface runoff. However, where intensity and seasonality of precipitation may create surface runoff in irrigated or nonirrigated areas, there are opportunities to investigate surface runoff losses. Nelson (1973), Dunnigan et al. (1976), and Kilmer et al. (1974) have all shown relatively small nitrogen losses from surface runoff as a result of precipitation in areas with varying slopes and amounts of precipitation. Losses for a variety of cropping and cultural systems, sources of nitrogen, rates of application, and degree of incorporation vary from almost none to perhaps as high as 5 percent of the applied nitrogen due to surface runoff. Even in studies by Ayers and Branson (1973) and Miller and Wolfe (1977) where surface runoff and subsurface flows are considered as the aggregate agricultural return flows, the amount of nitrogen lost only amounts to 2 to 3 percent of the total soil nitrogen.

In terms of mass emission, the amount of nitrogen lost in surface waters is relatively small, but its concentration is of most concern in the runoff or return flows. Carter et al. (1971) found that the concentration of nitrate-N in the surface runoff averaged 0.12 ppm. This was approximately a 100 times lower concentration than found in subsurface drainage waters. Busch et al. (1975) found mass emissions of nitrate-N in surface runoff water ranging from practically nil to

0.6 kg/ha for different cropping systems under surface irrigation. The majority of the observed losses were below 0.3 kg/ha based on the amount of surface runoff recorded and concentrations ranging from 0.05 to 0.7 ppm of nitrate-N.

As pointed out by Tanji et al. (1976), in some cases the quality of tailwater may actually be improved by irrigation use through the removal of suspended solids or reduction in nutrient concentration of the irrigation water. In a study in Glenn-Colusa Irrigation District, Tanji et al. (1976) showed that water discharges from the district had nitrate concentrations ranging from approximately 0.8 to 2.6 ppm. The diverted water had 1 ppm nitrogen. They concluded that there was relatively little change in the nitrogen concentration in the tailwater compared to the diverted supply water.

SIMULATION/PREDICTION MODELING OF NITROGEN SOURCES AND SINKS

As pointed out by Tanji and Gupta (1978), a simulation model attempts to forecast how a system will behave without actually making measurements in the physical system. The model is expected to reflect all the processes and relationships of the system. It would be desirable for the model to be universally applicable to a wide range of conditions with a minimum amount of complexity. They identified four types of models; a lumped-parameter or system, a distributed-parameter system, a stochastic, and a deterministic model. The lumped-parameter model is an input/output approach where the processes and mechanics of change are not considered. It is commonly referred to as a "black box analysis." The distributed-parameter model examines the internal processes of the system separately and then weights their effects upon each other in order to arrive at an output. The stochastic model is applied to a system in which the activity at any one time is not explicitly known but is addressed as a probabilistic event. The deterministic model examines the probability of the outcome based upon definite cause-and-effect relationships. Depending upon the availability of information and the extent of elucidation of the processes involved, a simulation/prediction model might also consist of subroutine or submodels, which may employ all of the different types of models identified.

The degree of sophistication of the model, then, as pointed by Davidson et al. (1978), will depend upon the understanding of the system, the available data base for validation, and the intended application of the model predictions. An excellent review of models currently employed in predicting water transport, nitrogen transformations, and nitrogen and water transport in a soil–plant–water system was provided by Tanji and Gupta (1978) with critiques in the same review by Davidson et al. (1978), and by Frissel and van Veen (1978). The review covers the overall assumption of the model, the degree of sophistication, and the specific processes included in the model.

In Table 5-9 some simulation models are listed that have been used to predict soil water transport, nitrogen transformations, and nitrogen transport. Along with their authors some information is given on their application, availability, and the processes considered. Although a process or factor may be indicated as one

Table 5-9. Selected nitrogen simulation models.

References	Brief description
Dutt et al., 1972	Simulates moisture flow, N uptake, residual soil organic matter, and nitrates in leachate.
Shaffer et al., 1977	Updated version of Dutt et al. (1972) for tile drained croplands, denitrification and transition state nitrification added.
Mehran and Tanji, 1974	Simulates first-order kinetics for nitrification, denitrification, mineralization, immobilization, and plant uptake and reversible first-order kinetics for ammonium-ion exchange.
Tanji et al., 1981	Updated version of Mehran and Tanji (1974) to include coupled water flow with N transformations and transport. Model includes daily mass balance of water and N.
Hagin and Amberger, 1974	Simulates nitrification, oxygen transfer in soils, denitrification, and evapotranspiration.
Frissel and van Veen, 1978	Simulates mineralization, nitrification, denitrification, ammonia volatilization, ammonium clay fixation, and nitrate leaching.
Tiliotson et al., 1980	Simulates water flow, N-transformations, urea hydrolysis, ammonia volatilization, and plant uptake.
Rao et al., 1981	Simulates water flow, N-transformations, water extraction, N uptake, multi-layered soils, and N behavior in cropped lands receiving crop residues and animal manures.
Selim and Iskander, 1981	Simulates N behavior in land treatment systems, physical, chemical, and biological process of N transformations to estimate the leachate concentration of nitrate.
Carbon et al. 1991	Simulates mineralization, nitrification, ammonium adsorption-desorption, immobilization of N from ammonium and nitrate, N uptake by plants, leaching of nitrate, water content and movement.
Wagenet and Hutson 1989	Simulates plant uptake, mineralization, immobilization, leaching, denitrification, volatilization, water and heat transport and evapotranspiration.

Continued--

- Table 5-9. (Continued)

Johnsson et al., 1987	Simulates plant uptake, mineralization, immobilization, leaching, denitrification, and water and heat flow.
Rijtema et al., 1990	Simulates mineralization, immobilization, uptake by plants, denitrification, soil-water dynamics, nitrogen leaching, volatilization, ammonia adsorption-desorption, nitrification and dissolution.
Lafolie, 1991	Simulates ammonium adsorption-desorption, plant uptake, nitrification, mineralization, denitrification, and leaching.
Verrecken et al., 1990	Simulates nitrification, denitrification, volatilization, mineralization, immobilization, and ammonium adsorption- desorption.
Hansen et al., 1990	Simulates soil water dynamics, soil temperature, soil nitrogen with organic matter dynamics, net mineralization of nitrogen, nitrification, denitrification, nitrogen uptake by plants and nitrogen leaching.
Groot and de Willigen, 1991	Simulates N mineralization, immobilization, transport, plant uptake, crop growth, and soil water dynamics.
Kersebaum and Richter, 1991	Simulates N mineralization, transport of water and nitrate, plant growth and nitrate uptake.

of the aspects of a model, it does not indicate the degree of sophistication employed. For example, plant uptake may be treated in one model as a process involving mass flow, diffusion, root growth rate, effective root surface area, and transpiration. In another model it may consider only transpiration and average nitrate concentration in the soil solution. This points out how critical it is to obtain a copy of the author's paper to become familiar with the program and its assumptions before one can evaluate the model's usefulness in predicting soil nitrogen or water movement in a given system. Thus far, the database is not available for any serious attempts at developing a model incorporating spatial variability within that field. A fairly basic assumption in each of the models is the uniform application of water and nitrogen. As Tanji and Gupta (1978) point out, the placement of fertilizer in bands and the inherent spatial variability in field sampling procedures are aspects which one-dimensional models are incapable of handling. It does not appear that any models have been developed that adequately address the nonuniformity of water and nitrogen in a field.

NITROGEN FERTILIZER

In most irrigated soils, nitrogen is generally deficient. Exceptions may occur where there has been a long-term overfertilization, in peat soils, where the previous crop was a legume capable of fixing relatively large quantities of nitrogen and returned to the soil, or where legumes are grown. In most situations nitrogen must be supplied as fertilizer in order to produce optimum or maximum yields. There has been an increase in fertilizer nitrogen consumption as native soil nitrogen supplies have been depleted. McVickar (1967) reported that during the 100-year period 1850 through 1949 the total amount of nitrogen fertilizer sold in the United States was only slightly greater than the amount sold during the year in 1950 alone. For the same period, it had been estimated by George Stanford, as cited by Veits and Aldrich (1973), that approximately 1.6 billion metric tons of organic nitrogen were lost from cultivated soils of the United States. Assuming an average of 162 million hectares of land under cultivation during the same period, this would amount to approximately 196 kg/ha of N lost or removed. How much of the loss was a result of erosion, denitrification, leaching, and crop removal is not known.

Apparently, the amount of nitrogen indicated above was adequate for the yield potentials obtainable during that period. The adequacy of the mineralizable nitrogen for crop production is evident from examination of the late 19th century and early 20th century experiment station bulletins. It was frequently reported that no measurable increases in yield had resulted from nitrogen fertilizer application.

Nitrogen fertilizer consumption levels for 1982 are shown in Table 5-10 for major agricultural states in the United States. States with extensive and intensive crop production, especially winter vegetable production for the fresh market, have the highest intensity of fertilizer use per unit of land area. The intensity of use data are somewhat misleading in that not all cropland is fertilized in a given year. As a consequence of this the actual intensity of use is expected to be greater. This is certainly true for high cash value crops.

Nitrogen fertilizer consumption for the years 1968, 1978, and 1988 is shown in Table 5-11 for nine regions of the United States. While the rate of growth for nitrogen fertilizer consumption has slowed in the second decade, large changes in the nitrogen fertilizer materials used between 1970 and 1990 have occurred (Fig. 5-10). Ammonium nitrate use has declined, whereas use of anhy-

Table 5-10. Nitrogen fertilizer use in major agricultural states*, 1982.

State	Nitrogen Consumption (metric tons x 1000)	Intensity of Use (kg\ha)
Alabama	111	74
Arizona	72	149
Arkansas	179	55
California	432	112
Colorado	26	34
Florida	209	174
Georgia	194	89
Idaho	170	71
Illinois	707	72
Indiana	376	73
Iowa	671	66
Kansas	477	41
Kentucky	144	67
Louisiana	106	52
Maryland	47	73
Michigan	186	59
Minnesota	368	43
Mississippi	137	54
Missouri	276	50
Montana	107	17
Nebraska	523	61
New Mexico	25	41
New York	74	40
North Carolina	188	90
North Dakota	197	17
Ohio	340	77
Oklahoma	224	52
Oregon	131	75
Pennsylvania	71	39
South Carolina	71	62
South Dakota	72	10
Tennessee	106	53
Texas	594	54
Utah	26	45
Virginia	73	58
Washington	210	67
Wisconsin	174	39
Wyoming	37	40
TOTAL	8,232	Avg.63

Source: U.S.D.A. 1983.

* States with at least 0.4 million hectares (1 million acres) of cropland used for crops in 1982.

drous ammonia, urea, and nitrogen solutions has increased with urea use having the greatest rate of increase.

In the western United States irrigated areas account for a relatively large proportion of the total fertilizer nitrogen use by those states. This is assumed to reflect the relatively higher yield potential achievable for warmer growing conditions and where water input is not subject to climatic variations. The data also show a reduction in rate of growth of the amount of nitrogen fertilizer sold in the United States by the end of this period. This is attributed to the relatively small change in total area under cultivation and high percentage of cropped areas receiving adequate nitrogen fertilizer as well as the impact of recent legislation establishing "best management practices" for applying nitrogen fertilizers.

Table 5-11.   Nitrogen fertilizer consumption in the United States by region.

REGION	1968	1978	1988
	----------------------------------(tons)----------------------------		
New England	40,392	43,428	37,434
Middle Atlantic	230,027	290,220	233,864
South Atlantic	722,624	937,033	756,414
East North Central	1,319,714	2,074,220	2,181,696
West North Central	2,139,828	3,364,240	3,598,269
East South Central	442,435	616,826	605,791
West South Central	880,400	1,179,926	1,570,540
Mountain	323,919	534,893	583,083
Pacific	637,245	878,477	909,351
Total U.S	6,737,583	9,964,619	10,505,400

Source: Berry and Hargett, 1989.

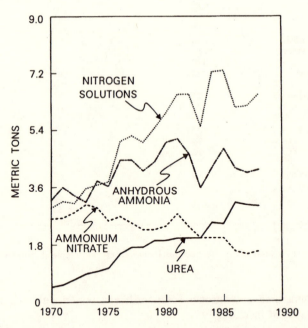

Figure 5-10   Changes in nitrogen fertilizer materials used in the United States from 1970 to 1988. (Berry and Hargett, 1989.)

Although the United States may be approaching relatively low growth with respect to consumption of nitrogen fertilizer, the same is not true in terms of world fertilizer needs. This fertilizer demand will result from improvement in cultural practices and adaptation of crops with higher genetic yield capability to various agricultural areas in the world. The technology for such improvements already exists and should provide several-fold increases in yield as socioeconomic conditions reach the point where the technology can be implemented.

The dramatic increases projected for worldwide fertilizer nitrogen consumption come at a time when the energy feedstocks for nitrogen manufacture are being depleted and costs are high. Approximately 38,000 cubic feet of natural gas are required to produce one ton of ammonia gas. Once nitrogen has been industrially fixed in the form of ammonia, it can be used as the raw material for the manufacture of ammonium, nitrate, urea, and other nitrogen fertilizers. A comparison of the cost of ammonia production using four different feedstocks is shown in Fig. 5-11. These data (Sharratt, 1976) indicate the dilemma faced by nitrogen producers. Most of the world capacity is based on natural gas and naphtha feedstocks. If the cost of current feedstocks become exorbitant and the switch to other energy feedstocks is necessitated, it can be accomplished technically. In either case, the cost of nitrogen will be substantially greater than at present.

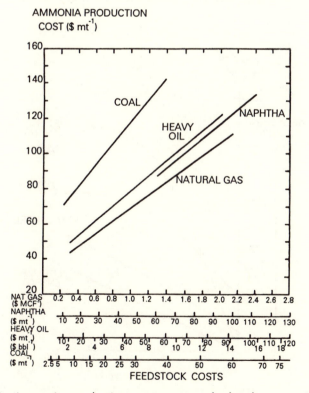

Figure 5-11 Ammonia production costs versus feedstock costs, 100 metric tons/day. (Sharratt, 1976.)

Another aspect of the energy situation with respect to the amounts required for fertilizer nitrogen manufacture is the amount of energy used in relation to the total energy consumed in the United States. Sartain (1976) indicates that the fertilizer industry uses about 12 percent of the energy consumed in agriculture, which amounts to about 1.5 percent of the total U.S. energy consumption. This agrees with information developed in a report of the energy requirements for California agriculture by Corvinka et al. (1974). California agriculture used about 5 percent of the energy consumed in the state, with about 15 percent of that used for fertilizer production, transportation, and field application. Consequently, only about 0.75 percent of the total energy consumed in the state was being used to supply fertilizer for agriculture. The relationships between the yield of grain sorghum and energy inputs in the form of fertilizer and other cultural practices are shown in Fig. 5-12. Almost three times as much grain yield was obtained on the same land area when an adequate level of nitrogen fertilizer was applied. Nitrogen applications in excess of that required for maximum yield lead to substantially reduced net energy return. However, it should not be assumed that a net energy return from a crop is essential to justify fertilizer input. Many crops are produced for reasons other than their caloric value. For example, vitamins, minerals, and quality of diet are important considerations in determining the type of crops to be produced. Nevertheless, Fig. 5-12 does indicate that energy efficiency is closely related to maximum

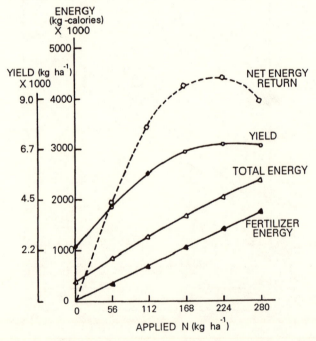

Figure 5-12 The relationship between use of nitrogen fertilizer and yield of grain sorghum, energy inputs, and energy returns. (Rauschkolb, 1974a.)

Table 5-12. Principal manufactured fertilizer sources of nitrogen.

Fertilizer	Nitrogen form in percent				Relative cost[a]	Physical state	Net[b] acidity (A) alkalinity (B)	Salt[c] index	Chemical formula	Other nutrients
	Ammonia	Nitrate	Urea	Total						
Ammonia	82			82	1.00	gas	A-147	47	NH_3	
Ammonium nitrate	17	17		34	2.10	solid	A-62	105	NH_4NO_3	
Ammonium sulfate	21			21	1.89	solid	A-110	69	$(NH_4)_2SO_4$	24% S
Aqua ammonia	20			20	1.19	liquid	A-36	--[d]	NH_4OH	
Calcium nitrate		15.5		15.5	2.59	solid	B-20	53	$Ca(NO_3)_2$	21% Ca
Diammonium phosphate	16-18			16-18	--[e]	solid or liquid	A-70	34	$(NH_4)_2HPO_4$	48 - 48% as P_2O_6
Mono-ammonium phosphate	11			11	--	solid or liquid	A-58	30	$NH_4H_2PO_4$	48% P as P_2O_6
Potassium nitrate		13.5		13.5	--	solid	A-23	74	KNO_3	46% K as K_2O
Urea			46	46	1.67	solid	A-71	75	$CO(NH_2)_2$	
Urea-ammonium nitrate	8	8	16	32	1.69	liquid	A-57	--	$CO(NH_2)_2$ plus NH_4NO_3	

[a] Relative cost is calculated by dividing the dealer cost per pound of N for each fertilizer by the cost per pound of N for ammonia. Actual costs will vary with dealer, transportation costs, and quantity purchased. The relative costs may vary from location to location.

[b] Acidity or alkalinity is based on the equivalence to pure limestone ($CaCO_3$) in pounds of fertilizer material (Western Fertilizer Handbook, 1980).

[c] Relative salinity is based on the ratio of Osmotic pressures produced when a fertilizer material is compared with an equal weight of sodium nitrate multiplied by 100 (Hardesty, 1967).

[d] Data not available.

[e] Since these fertilizer materials contain nutrients other than nitrogen a relative cost for nitrogen could not be calculated.

133

Table 5-13. Organic sources of nitrogen[e].

Source	Range of composition[b]	Typical[b]	Availability in year of application[c]
	---------------------------%---------------------------		
Animal manures:			
Beef	1.5 - 7.8	2.0	Fresh 75[d]
Dairy	1.6 - 4.0	3.3	Corral 40[d]
Horse	1.7 - 3.0	2.5	40
Poultry	2.1 - 5.9	3.5	90[d]
Sheep	3.0 - 4.5	3.5	30
Swine	2.0 - 5.6	2.8	60
Food processing waste:			
Peach (halves, skins, & seeds)	0.6 - 0.9	0.7	30
Peach-Pear (halves, skins & seeds)	0.8 - 1.0	0.9	40
Pear (halves, skins, and core)	---	1.4	50
Potato (waste water),ppm	32 - 133	55.0	80
Tomato (skins, juice, & seeds)	1.8 - 3.2	2.8	60
Winery pomace	1.0 - 2.0	1.5	40
Municipal sewage:			
Sludge	1.0 - 6.0	2.0	35[d]
Effluent, ppm	15 - 40	25	75
Miscellaneous other:			
Alfalfa hay	2.2 - 3.0	2.5	70
Bat guano	9.7 - 13.8	13.0	80
Cotton gin trash	---	0.7	40
Dried blood	---	13.0	80
Fish scraps	8.9 - 10.4	9.2	80
Grain straw	0.4 - 0.8	0.2	30

[a] Table was constructed from data reported by: Abbott, 1975; Azevedo and Stout, 1974; Meek et al., 1975; Menzies and Chaney, 1974; Pratt et al., 1973; Rauschkolb et al., 1975; Reed et al., 1973; Smith et al., 1976; Tisdale and Nelson, 1966; and the Western Fertilizer Handbook, 1980.

[b] Except where otherwise indicated values are reported in percent on a dry weight basis.

[c] The estimated availabilities are approximate values based upon nitrogen response studies, mineralization studies, and/or nitrogen content and type of nitrogenous compounds in various materials.

[d] Values reported by Pratt et al., 1973.

[e] No value for the range in N content reported in the reference source.

crop yield. Regardless of the amount of energy return per unit of energy input, the highest relative energy efficiency occurs when adequate fertilizer is applied to obtain maximum yield.

Some of the principal nitrogen fertilizer sources are presented in Table 5-12. Such factors as relative costs and the physical and chemical characteristics of materials provide a basis for selecting fertilizer materials most appropriate for the cultural practices and conditions under which may be used.

Current emphasis on conservation of resources has generated renewed interest in organic materials as sources of nitrogen for plants. In Table 5-13 are presented different types of organic nitrogen sources along with their range in composition and estimated nitrogen available. In some instances, relatively good information is available for establishing a nitrogen availability value based on the nitrogen mineralization in the year of application. However, it is not possible for estimates of nitrogen availability to be absolute since treatment, storage, handling, and climatic factors interact to determine the actual nitrogen availability of the material. These estimates are intended to provide a guide for determining appropriate rates of application so that nitrogen excesses can be avoided.

Procedures outlined by Pratt et al. (1973) utilize the decay series to adjust the rate of organic waste application. Figures 5-13 and 5-14 indicate how a decay

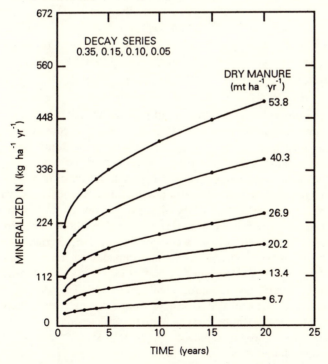

Figure 5-13 Yearly mineralization rate in relation to time for various constant rates of corral manure having 25 percent water and 1.5 percent nitrogen on a dry weight basis. (Pratt et al., 1973.)

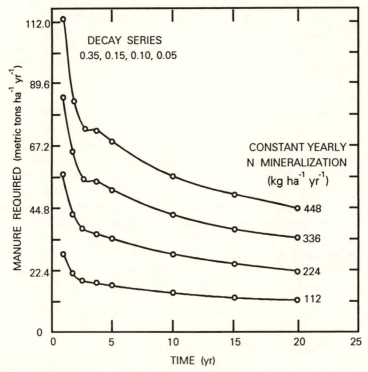

Figure 5-14 Yearly rates of application of manure containing 25 percent water and 1.5 percent nitrogen on a dry weight basis, required to maintain various constant yearly rates of nitrogen mineralization. (Pratt et al., 1973.)

series may be utilized to adjust the rate of dry manure applications in order to achieve different levels of available nitrogen. This can be accomplished by either maintaining a constant level of manure application resulting in increasing amounts of mineralized N with time, or by adjusting the amount of manure applied with time to maintain a relatively constant amount of mineralized N. In Table 5-14 are presented the decay series for different materials, where the numbers indicate the percentage of the nitrogen mineralized for any given year. Two decay series are presented: a slower rate, which might be expected in areas where colder climates restrict the decomposition of the material, and a faster rate, which would be applicable to warmer climates where soils are seldom subject to freezing. As pointed out by Meek et al. (1975), climate is important in determining annual rate of nitrogen mineralization. They have provided a method for calculating annual degree-days from commonly available climatological data. Such value can be used as a guide for determining where the slow or fast decay series may be employed. Figure 5-15 shows the climatic regions for the western United States.

In determining amounts of organic waste to use, nitrogen content and concentrations of potentially toxic or harmful constituents are important. Because of

Table 5-14.   Decay series for predicting the rate of nitrogen mineralization for various organic wastes used as nitrogen fertilizers.

Material		Years After Application Percent of mineralized N[a]						
		1	2	3	4	5	6	7
Chicken manure	S[b]	90	10	7.5	5	4	3	
	F	90	10	5				
Fresh bovine,	S	75	15	10	7.5	5	4	3
3.5% N	F	75	15	10	5			
Dry corral manure,	S	40	25	6	3			
2.5% N	F	40	25	6				
Dry corral manure,	S	35	15	10	7.5	5	4	
1.5% N	F	35	15	10	5	5	4	
Dry corral manure,	S	20	10	7.5	5	4	3	
1% N	F	20	10	5				
Liquid sludge,	S	35	10	6	5	4	3	
2.5% N	F	35	10	5				

Source: Pratt et al., 1973.

[a] In the first year after application a portion of the total nitrogen in the material is mineralized; the second year after application a portion of the remaining nitrogen is mineralized; and so on until the rate of mineralization of the remaining nitrogen is approximately the same as the soil humus.

[b] The S = slow and F = fast rates of mineralization to reflect the climatic influence on decomposition of the organic material, S for colder climates, F for warmer climates where the soils rarely if ever freeze.

Table 5-15.   Some slow release fertilizer nitrogen sources.

Fertilizer	Nitrogen content	Other nutrient	Method of controlling release
	---%---		
Isobutylidene-diurea (IBDU)	31	N/A	Particle size regulating solubility
Osmocote R[a]	14 to 18	5 to 14 P_2O_5 11 to 14 K_2O	Resin coating of fertilizer material.
Sulphur coated urea	32 to 37	16 to 20 S	Coating with molten sulfur.
Urea formaldehyde	35 to 38	N/A	Particle size regulating solubility and rate of decomposition.

[a] Trade name of the material.

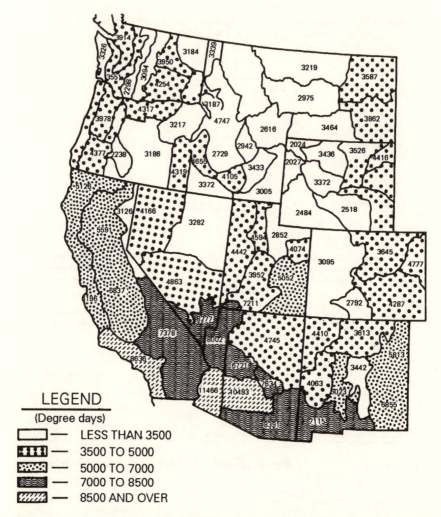

Figure 5-15 Degree days in the western United States. Groupings have been made for less than 3500, 3500 to 5000, 5000 to 7000, 7000 to 8500, and over 8500 (see legend). These divisions are arbitrary. Degree days were calculated from summation of the year for average monthly temperature (°F) minus 40°F multiplied by the number of days per month. (Meek et al., 1975.)

the relatively high salt content of manure, it is often desirable to utilize the manure in as fresh a form as possible. In this state nitrogen content is at its highest, thus the amount that must be applied to provide a required amount of nitrogen for plants is the lowest. The net effect is a higher nitrogen availability and a lower salt load. For food processing wastes, the principal controlling factor is the nitrogen concentration of the material and its availability. Where processing wastewaters are being utilized, the ability of the soil to infiltrate the amount of water applied is also a factor. If the nitrogen concentration is low, then the evapotranspiration needs of the crop will generally determine the amount of the wastewater that can be applied. Where the nitrogen concentration is high, then the nitrogen loading rate becomes the limiting factor.

For municipal sewage sludge, trace metals are also present that can impose another limitation to the amount of this material that can be applied to soils. This is especially true in areas where a variety of crops are grown for human consumption. Where land can be dedicated to seed crops that have relatively low uptake and accumulation of trace elements, much higher loading rates can be tolerated. However, land may have to be dedicated to a particular use when loading rates of trace elements are greater than can be tolerated by crops grown for human consumption.

Another technique employed in regulating the nutrient supply to plants has been the use of controlled-release fertilizers. Under conditions where there is a high potential for leaching or denitrification losses, controlling the release of nitrogen to the soil may provide a means for increasing the efficiency of nitrogen fertilizer use. Some of the common forms of slow release materials with their respective nitrogen content are shown in Table 5-15. By means of controlling water solubility or rate of decomposition, the rates of nitrogen release can be controlled to an appreciable extent.

6

Environment

Climate is an integral aspect of environment and in one sense the words are synonymous. Plant growth is easily recognized as being greatly influenced by the environment. Less readily apparent is the influence of climate on soils. However, it has long been recognized by soil scientists that climate is one of five basic soil-forming factors. In this chapter the climatic environment is examined from the standpoint of its influence on distribution of irrigated lands, nitrogen pools and transformation, and cropping patterns. The principal features of the environment that will be considered are precipitation and temperature.

PRECIPITATION

Precipitation in irrigated agricultural areas not only supplies water to plants to help meet their evapotranspiration needs, but also serves as a means of replenishing the surface and groundwater reservoirs that are the principal sources of water in irrigated agriculture. The normal monthly precipitation amount varies during the calendar year with periods of lower and higher precipitation. These distributions vary across the United States both in magnitude and in time within the year. These precipitation patterns determine the need for supplemental irrigation to assure adequate crop yields. In Florida, where annual rainfall averages 1400 mm, periods of drought of two or three weeks duration can cause serious crop damage and yield loss due to the low water-holding capacity of many sandy soils, especially entisols. The distribution of mean annual precipitation in the United States is given in Fig. 6-1. In the western United States, there are tremendous variations in topography, which create conditions that result in tremendous variation in the amount of annual precipitation. It is very common for annual precipitation to increase by more than 200 percent within a distance of 50 miles, as the terrain changes from a valley to a mountain summit (Gilford et al., 1967). Changes as large as 400 percent may occur in some cases within a distance of 10 miles. In the central, eastern, and southern areas of the United States, such extreme variation of precipitation with small changes in geographical location is not observed. One can also see from Fig. 6-1 that great differences occur in mean monthly precipitation for the irrigated areas in the west as com-

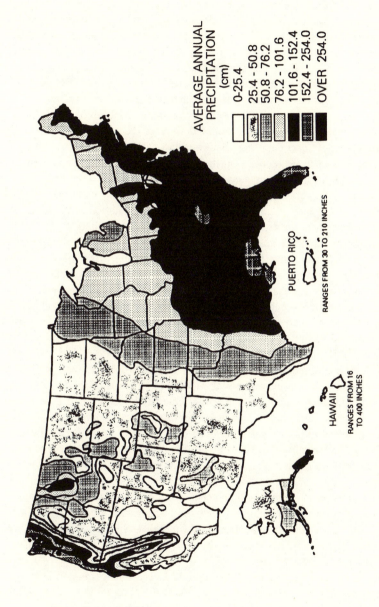

AVERAGE ANNUAL PRECIPITATION (cm)

0-25.4
25.4 - 50.8
50.8 - 76.2
76.2 - 101.6
101.6 - 152.4
152.4 - 254.0
OVER 254.0

PUERTO RICO
RANGES FROM 30 TO 210 INCHES

HAWAII
RANGES FROM 16 TO 400 INCHES

ALASKA

Figure 6-1 Average annual precipitation distribution in the United States. (U.S. Water Resource Council, 1968.)

pared to predominately rain-fed agricultural areas in the east. Total precipitation is much less in the western half of the contiguous United States.

Another distinctive feature is the distribution of precipitation along the west coast. In the winter months precipitation may be relatively high, while during the summer months it is extremely low. In most years for parts of Washington, Oregon, and California, the precipitation during June, July, and August at valley floor reporting stations is zero. This peculiarity in distribution of precipitation during the year accounts for the interspersion of rain-fed and irrigated agriculture in these areas. As one proceeds inland from the coast, the distribution pattern becomes more uniform. However, the total amount of precipitation is much less. In fact, the total amount of precipitation in these areas is considerably less than the potential for evapotranspiration, as indicated in Fig. 6-2.

The frequency of precipitation as well as the amount determines its relative effectiveness. Effective precipitation can be defined as the total amount received minus surface runoff and deep percolation. Seasonal irrigation water requirements are presumed to be reduced by the amount of effective precipitation. If all the precipitation came in one storm, runoff and percolation losses would be expected to be relatively large and effective rainfall would comprise only a small portion of the total. Consequently, effective precipitation is made up of numerous storms, each consisting of small quantities of precipitation; the result is an increase in the total amount of water available for evapotranspiration. According to Heermann and Shull (1970), the mitigating effect of increased evaporation from this type of precipitation pattern can cause cooling, thus reducing the quantity of water that plants can need for transpiration. The interaction of precipitation frequency and amount also has a significant impact on soil nitrogen pools. Any combination of frequency and amount resulting in a saturated soil condition enhances the potential for leaching and denitrification losses of nitrogen.

The influence of precipitation on nitrogen leaching may be seen in Table 6-1. The drainage volume increases with increasing amounts of rainfall as does the total amount of nitrogen being leached. One of the interesting aspects of these data is that nitrogen concentration decreases with increasing leaching volume. This is not surprising, in fact, it is to be expected, but it does point out the differences in magnitude of the problem one might expect from the impact of leached nitrate on groundwaters for rather arid versus relatively humid areas. In areas with higher rainfall, there is a greater amount of dilution water available so the concentration of nitrates in both surface and groundwater supplies would be expected to be lower. In central Washington, where rainfall is quite limited, Enfield (1975, personal communication, Carl Enfield, U.S.E.P.A. Robert S. Kerr Environmental Research Laboratory, Ada, Oklahoma) found little or no movement of water to the water table.

In addition to being an important contributing factor to leaching and denitrification, precipitation also contributes to the total soil nitrogen load. This influence is shown in Fig. 6-3 (McElroy et al., 1976). While nitrogen contributions of the magnitudes indicated are insignificant in terms of crop production under intensive management, the cumulative effect over years has been shown by Klemmedson and Jenny (1966) and Harridine and Jenny (1958) to be signifi-

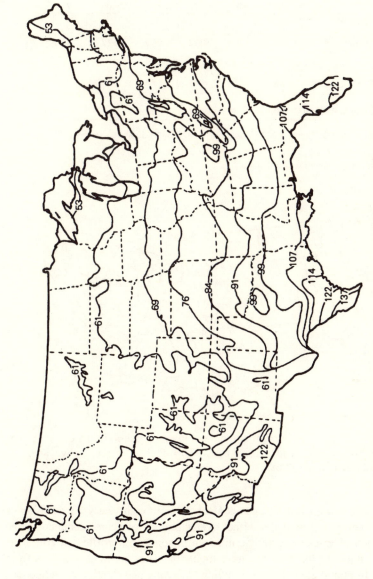

Figure 6-2 The average annual evapotranspiration potential (CM) for the contiguous United States. (Geraghty et al., 1973. Copyright © 1973 by the Water Information Center, Port Washington, NY. Reprinted by permission.)

Table 6-1.   Influence of rainfall amount on the quantity of water and nitrogen leached
              from lysimeters for soils with different textures*.

	Years					
	1964		1966		1968	
	Sandy	Loamy	Sandy	Loamy	Sandy	Loamy
Rainfall, mm	374		615		779	
Leached water, l/m²	95	93	312	309	527	497
Nitrogen in leached water: kg/ha	33	21	41	23	56	62
Drainage volume, ha-cm	9.5	9.3	31.2	30.9	52.7	49.7

Source: Adapted from Jung, 1972. Copyright © 1972 by W.Junk. Reprinted by
permission.

* Soil Depth in lysimeters was one meter.

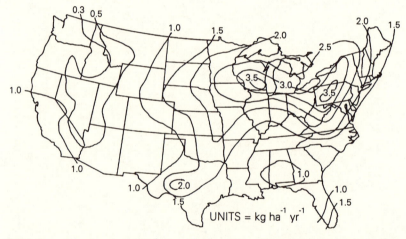

Figure 6-3 Nitrogen (ammonium and nitrate) in precipitation in the United States.
(Personal communication, J.H. Cravens, U.S.D.A.-Forest Service, Milwaukee, WI,
cited in McElroy et al., 1976.)

cantly correlated with soil nitrogen content. Even though yearly nitrogen inputs
are quite small, the cumulative effect over hundreds of years accounts for the
accumulation of so-called geologic nitrogen in some soil profiles. In arid areas
it is then subject to leaching as larger inputs of water are made, such as from
irrigation. In humid areas there is continual leaching and dilution of indigenous
soil nitrogen and yearly nitrogen inputs from precipitation.

Precipitation also has an influence on the magnitude of denitrification losses.
As soil moisture content is increased by either more frequent precipitation or
relatively large amounts of precipitation per storm, the soil has a tendency to
become saturated and denitrification losses increase. Reddy and Patrick (1976)
have shown the influence of the soil matrix cycling between aerobic and anoxic

conditions on the denitrification losses from soils. The nitrogen losses ranged from approximately 37 percent of the total nitrogen for the short-term cycles (6-hour, alternate aerobic–anaerobic) to approximately 26 percent for the long cycles (48-hour, alternate aerobic–anaerobic) when soil temperatures were maintained at approximately 30°C. The total nitrogen loss ranged from approximately 71 to 61 percent for the same aerobic–anaerobic cycling conditions. In this case, the cycling allowed for alternate nitrification and denitrification to occur. In another study by Pilot and Patrick (1972), soils ranging from loamy sand texture to silty clay loam texture were examined as to the influence of air-filled porosity and soil moisture tension on denitrification. At air-filled porosities of 11 percent for the loamy sand and 14 percent for the silty clay loam, aeration was adequate to prevent denitrification. This corresponded to soil-moisture tensions of between 20 and 30 cm tension, the lower value for the coarser textured soil. Consequently, one can see that frequency and amount of precipitation, as it influences moisture content of soils, would have a significant impact on the magnitude of denitrification that can occur in a soil. And as Pilot and Patrick (1972) indicate, the soil-moisture condition necessary to initiate denitrification is specific for each soil.

When one imposes the additional variable of precipitation on widely varying soil types, then it is easily understood why much difficulty is encountered in managing nitrogen in irrigated areas. In addition to the direct effects of soil moisture on the soil nitrogen pool and plant growth, precipitation also has an effect on soil temperature. Poorly drained soils are generally much colder than are well-drained soils because of the greater amount of radiant energy required to heat the water in them. This is the result of the greater specific heat of water than of soil minerals (Thorne and Peterson, 1954). Observations by numerous others have shown the cooling effect of soil moisture. The effect may be beneficial or adverse depending upon the time of the year and the effect on soil drying, germination, and plant growth.

TEMPERATURE

The influence of temperature on chemical and biological reactions has been examined by numerous investigators. It is a widely recognized phenomenon that the rate of most chemical reactions doubles with every 10°C temperature. In the metabolism of plants and soil microorganisms, temperature governs the rates of chemical reactions and enzymatic activity. In instances where enzymes may exist outside the cell, such as urease, the influence of temperature is the same. High temperatures can cause a denaturization of enzymes and lower temperatures can result in inactivation or denaturization of the enzyme. According to Treshow (1970), biochemical reactions have been shown to increase with a doubling of rate of most reactions with every 10°C temperature increase from 10°C to about 30°C. He indicates that below about 10°C enzyme activity is minimum. For example, photosynthesis can be considered negligible below 10°C. Even though photosynthesis is principally a photochemical process, it still depends upon enzymatic activity.

Plant Adaptation

The influence of temperature on plants and microbial reactions is much more complex than might be inferred from a simple examination of the influence of temperature on reaction rates. Plants have adapted to a wide range of temperature regimes, such as the Arctic Tundra, the Equatorial Forest, and deserts. Diurnal and seasonal variations in temperatures also affect plant growth. For example, Went (1957) indicated that tomato, potato, and pepper plants are known to develop best when moderate daytime temperatures are associated with cool nights. Seasonal variation is important for many agronomic and horticultural crops. For example, several varieties of wheat require vernalization to induce flowering. Similar effects on flowering have been found for tree crops such as apples, peaches, apricots, pistachios, and pecans. In general, plants are known to grow over a wide range of temperatures, with optimum growth occurring in a fairly narrow temperature range. The temperature for which optimum growth occurs for different species may tend to overlap, as indicated in Fig. 6-4. The temperature extremes at which the plant can still grow at a relatively good rate differ considerably for wheat and corn (Hatfield, Jerry, 1977, personal communication, formerly Bioclimatologist at the University of California, Davis, currently Director of the Tilth Laboratory, USDA-ARS, Ames, Iowa).

The length of season during which temperatures are satisfactory for plant growth is also critical. Plants are known to be adapted to growing seasons ranging from as little as 90 frost-free days to year-round growing conditions. As indicated by Loomis (1976), corn requires at least 120 days during which the temperature stays above 10°C. Plants have adapted to a wide variety of climatic and temperature variation so long as during part of the year the temperature and moisture regimes are satisfactory for plant growth. In a large part of the irrigated

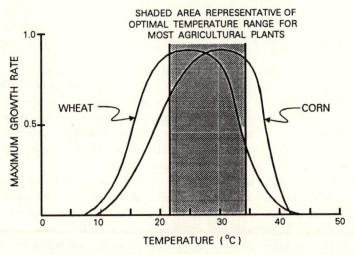

Figure 6-4 Relationship between temperature and growth rate for corn (*Zea mays*) and wheat (*Triticum aestivum*). (Personal communication, J. Hatfield, 1977.)

western United States the temperature regime is such that plants may be grown almost year-round so long as evapotranspiration needs are met by rainfall or irrigation. Multiple cropping of the same land is practiced in those areas by selecting crops that are adapted to the temperature regime of a given season. In the so called "sun belt" of the United States this practice is widely employed.

Temperature and relative humidity have a marked effect on evapotranspiration. In the Pacific Northwest and parts of California winter rainfall may provide most of the evapotranspiration needs of plants grown during the winter months (see Figs. 6-1 and 6-2). However, in most of the irrigated areas winter irrigation is required to meet the evapotranspiration needs of plants that are adapted to the temperature regimes of those areas. Recognition of the seasonality of irrigation requirements is important from the standpoint of water resource management.

Nitrogen Absorption

In addition to temperature effects on plant selection, cropping patterns, and water use, temperature also has a marked effect on nitrogen uptake. In studies examining the influence of different root and air temperature on uptake of nitrogen by lettuce plants, Frota and Tucker (1972) found that ambient air temperature, as well as root temperature, had a significant influence on the uptake of nitrate and ammonium forms of nitrogen. There was little difference in uptake of either form at 8°C and 23°C, but at 13°C and 18°C the rate of ammonium absorption was greater than the rate of nitrate absorption. When the initial root temperature was 2°C and the root was subsequently warmed to 23°C there was a slightly greater uptake of nitrate nitrogen during the 10-hour absorption period. At much lower ambient air temperatures and root temperatures, the absorption of ammonium was favored over that of nitrate. Although nitrate uptake was much reduced at lower temperatures, a greater amount of the absorbed nitrate was translocated to the tops of the plants than of ammonium for the ammonium-treated plants. In any case, both ammonium and nitrate absorption were very much reduced by the low temperatures, although nitrate uptake seems to be more restrictive at low temperatures.

In field studies conducted by Gardner and Pew (1972, 1974) they found that uptake and storage of nitrogen by lettuce plants was markedly reduced when mean weekly ambient air temperatures fell below 55°F (approximately 13°C). Using different nitrogen fertility levels they found that a reduction in the nitrate nitrogen level in the midrids of the outer leaves of the plant occurred at those low temperatures regardless of the nitrogen level in the soil or the source of nitrogen. Similar observations were made by Mayberry and Rauschkolb (1975) for lettuce grown in the Imperial Valley of California where the nitrate nitrogen concentration in midrib tissue of lettuce dropped at temperatures below 55°F (13°C). Historical average weekly temperatures for November through March indicate that there is a 2-week period during January in most years when the mean weekly temperature is below 55°F, which corresponds to what was observed. Consequently, it was suggested that no nitrogen fertilizer be applied just prior to or during this period. This is an example of how knowledge of the effect of temperature on nitrogen uptake can provide insight to the more efficient timing of nitrogen fertilizer applications.

Nitrogen Transformations

Biological transformations taking place in the soil profile are also affected by temperature. The influence of temperature on the change of ammonium nitrogen to nitrate was shown in Fig. 5-3, which was adapted from Parker and Larson (1962). After 22 days, the rate of nitrification was approximately the same for each soil temperature until the concentration of ammonium became a limiting factor. It appears that the effect of lower temperature was to delay the buildup of nitrifying organisms responsible for the transformation.

In other studies on the effects of temperature on nitrification, Frederick (1956) found the rate of ammonium transformation to nitrate ranged from 2 to 120 ppm per week as temperature increased from 2°C to 35°C (Fig. 6-5). In a soil with pH 5, the rate of nitrification for the same temperature range was from 0 to 30 ppm per week. As indicated previously (Fig. 5-3), it appears that the effect of temperature is to decrease the rate of buildup of microorganisms that can carry out nitrification. Similar temperature effects on nitrification were shown by Sokai (1959) where soils were held at 25°C until maximum nitrification occurred and then were subsequently maintained at a different temperature. Nitrification remained at a higher level than if the soil had been maintained at 5°C for the entire period. He concluded that nitrification may therefore occur at low temperatures if numbers of microorganisms are boosted.

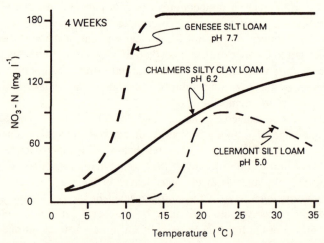

Figure 6-5 Influence of temperature on rate of ammonium transformation to nitrate for soils with different pH. Classification of soils are as follows: Chalmers (fine-silty, mixed, mesic typic Haplaquolls), Clermont (fine-silty, mixed, mesic typic Glossaqualfs), and Genesee (fine-loamy, mixed, nonacid, mesic typic Udifluvents.) (Frederick, 1956. Copyright © 1956 by the Soil Science Society of America. Reprinted by permission.)

Another nitrogen transformation influenced by temperature is denitrification. This process has been shown to occur at rapid rates under optimum conditions of temperature and microbial activity. Jordan et al. (1967) found that the rate of denitrification at 30°C ranged from 400 ppm nitrate-nitrogen denitrified per day to as little as 15 ppm per day, depending upon the microorganism involved. On the other hand, at temperatures below 10°C, Lance (1972) found that denitrification proceeded very slowly. Similar observations were made by Overrein (1968, 1969, 1971), who showed that about 25 percent of the nitrate was reduced at 4°C in 30 days, whereas approximately 95 percent of the nitrate was reduced at 12°C and 20°C during the same period. With the temperature held at 20°C almost 85 percent of the nitrate reduction occurred in the first 10 days. One might infer from this that the optimum temperature for denitrification is approximately 30°C. However, Broadbent and Clark (1965) reported in their review that they found the optimum temperature for denitrification to be in the range of 60°C to 65°C. These temperatures are generally destructive of other living organisms and are much higher than normal soil temperatures. Consequently, it appears that 30°C has been settled upon as a standard temperature for assaying denitrification potential of soils.

Temperature effects have also been observed on the rate of urea hydrolysis. Broadbent et al. (1958a) showed that 400 ppm nitrogen added as urea had completely hydrolyzed in 3 days at 24°C, whereas only 60 percent of the urea had hydrolyzed in the same period at 7°C. Even at the low temperature, assuming a linear rate of reaction, about 80 ppm of urea nitrogen was hydrolyzed each day. This rate is certainly adequate to hydrolyze urea at rates normally used for fertilization in a time period short enough to make nitrogen readily available for plant use and transformation to nitrate.

Physical Losses

In addition to biological activity, some of the physical losses of nitrogen are affected by temperature. Soil temperatures are generally well correlated with ambient air temperature. At the surface, soil temperature is higher than ambient air temperature; at about 5 cm depth, soil temperature is relatively close to ambient air temperature; but the peak soil temperature is slightly out of phase with the peak ambient air temperature. As one progresses further down the soil profile, the peak soil temperature lags further behind the peak ambient air temperature and the magnitude of the diurnal temperature change is dampened considerably. At 150 cm, the diurnal changes are no longer observable. Only seasonal changes are seen at this soil depth and such changes lag behind the seasonal changes in ambient air temperatures. The magnitude of seasonal soil temperature change depends upon the extremes of the ambient air temperature change.

Temperature also has been shown to have an effect on the viscosity of water, which is related to soil-water transport (Nielsen et al., 1970). Yet, it is difficult to ascribe a practical significance to this with respect to nitrate leaching because other soil characteristics may counter the reduction in soil-water viscosity as the temperature increases. These other characteristics include entrapped air and aggregate stability.

Volatilization of ammonium has been shown to be affected by temperature whether it is applied in water or on the soil surface. In a study of the amount of ammonia lost from water application through sprinklers, Henderson et al. (1955) found that the magnitude of the loss was highly dependent upon the concentration of ammonia in solution. However, at a given concentration the amount of loss increased with increasing temperature. The temperatures used in the study ranged from 20°C to 32°C, but the magnitude of the loss was relatively small because the rate of nitrogen application was low.

Greater amounts of ammonia volatilization losses have been reported where ammonium nitrate, ammonium phosphate, ammonium sulfate, or urea have been surface-applied. In studies by Shankaracharya and Mexta (1969) on a soil with a slightly alkaline pH, they found the ammonium losses from surface application of urea increased from about 2.5 percent at 20°C to slightly above 30 percent over 45°C at a relatively high rate of nitrogen application. Other studies have shown higher rates of volatilization losses from urea at 30°C on a soil with a similar pH and at a comparable rate of application. Figure 5-8 showed the influence of temperature on the magnitude of ammonium volatilization losses from ammonium nitrate and from ammonium sulfate or diammonium phosphate (Fenn and Kissel, 1974). There does not appear to be any effect of nitrogen rate on the magnitude of the loss when ammonium is applied as ammonium nitrate. For the ammonium sulfate and phosphate sources, there appears to be both a rate and temperature effect on the magnitude of the ammonia volatilization loss. Temperature appears to have little effect at either very low or very high rates of application. At intermediate rates of application the magnitude of the loss increases with increasing temperature. The effect of volatilization losses is to reduce the efficiency of nitrogen fertilizer use.

SIMULATION OF ENVIRONMENTAL VARIABLES

Prediction of climate in a given locale is at best a developing art, according to Jerry Hatfield (personal communication). Studies have been conducted only in a few limited locations to determine the probability of precipitation and temperature for a given day or week. This information is then used to extrapolate the occurrence of these events in nearby areas where they have not been tested. Techniques for determining probability models are statistical, with the most common approach being the use of Monte Carlo estimation methods similar to those outlined by Hahn and Shopiro (1967).

Gilford et al. (1967) show rainfall probability for selected stations in Colorado and indicate that precipitation probabilities also can be developed for other areas in consultation with the state agricultural experiment station and the state climatologist for the ESSA Weather Bureau. Both Gilford et al. (1967) and Hatfield (personal communication, 1977) indicate that the largest problem encountered in simulating climate is the length of record available, the quality of the record, and the uncertainties of extrapolating information from reporting stations to nearby areas. Because of topographical changes over rather short distances, adjacent areas may experience entirely different temperature and precipitation regimes.

AUTOMATED CLIMATE NETWORKS

With the advent of improved electronic technology there has been a major change in the type of instrumentation available for accumulating and reporting climatic data. Significant among these changes has been the development of automated weather data collection stations that can be remotely located and transmit data to a computer for processing. These stations may be referred to as "real-time" weather stations since the data is collected at very short intervals, sometimes every few seconds. While technology exists for these stations to transmit data by low-frequency radio, infrared, and microwave via satellite, the most reliable and preferred method of data collection is by having the station hard-wired to a telephone line and using a modem, telephone, and computer to access the data. Similar stations are used by the National Weather Service of the National Oceanic and Atmospheric Administration and the Federal Aviation Administration. These data are not readily available to the public on-line nor do they include the types of data required to make the calculations needed for some of the biological and physical models that are important to the agriculturalist or environmentalist.

Several states, through the College of Agriculture at their Land-Grant University, have installed weather collection and reporting networks that serve the state or larger area. Arizona has a network called AZMET (Arizona Meteorological Network) that has 17 stations located in agriculturally important areas of the state. The central computer is located on the campus of the University of Arizona and is operated by the College of Agriculture. The public can access the information on-line by using a modem, telephone, and computer once one has obtained an identification number. There is no charge for the information. California has a network called CIMIS (California Irrigation Management Information System) that has 50 stations located in agriculturally important areas of the state. The network was established by a grant from the California Department of Water Resources (DWR) to the University of California. Once the system efficacy was established the central computer and network reverted to DWR for continued operation. The information is available on-line to a user by modem, telephone, and computer once one has obtained an identification number. There is no charge for the information. Nebraska has a weather network as part of the AGNET (Agricultural Network) system operated by the College of Agriculture at the University of Nebraska. The AGNET system has other states as collaborating members to access a wide range of information available through AGNET. The weather network portion of AGNET is being used by the Colleges of Agriculture at Iowa State University and South Dakota State University. There are over 100 stations in the three states located in agriculturally important areas of the states. The information is available on-line by modem, telephone, and computer. There is a fee for obtaining an identification number to access AGNET for which one is able to use all the services including the weather data.

The weather data is collected at the site using a data logger to record the data at very short time intervals (varying from a few seconds to a few minutes). A central computer queries the data logger at all the stations in the network, usually once each 24 hours, and downloads the data for processing. The data collection and processing procedures may take a few hours depending upon the number of stations and the calculations made from the data. Once the processing is completed the information is available to a user who has a modem, telephone, and computer to access the data.

Typically the data collected is ambient air temperature (this is used to determine maximum and minimum air temperature), relative humidity, precipitation, solar radiation, wind speed, and wind direction. Additional information such as pan evaporation and soil temperature at different depths may be obtained, depending upon particular needs of the users of this data for predictive models or other purposes. The data may be collected at very short intervals but only reported as hourly averages. Frequently the network central computer calculates from these such factors as potential evapotranspiration, growing degree-days, and hours of chilling. All of these are important factors in plant and pest response to changes in the environment.

PART II
MANAGEMENT VARIABLES

7

Placement

The placement of nitrogen fertilizers is important in governing the efficiency of nitrogen fertilizer use. By utilizing a variety of placement techniques, it is possible to control such factors as volatilization losses, leaching of nitrogen below the root zone, uptake of nitrogen by the plant, and nitrogen transformation such as denitrification, mineralization, nitrification, and immobilization. The influence of placement on nitrogen utilization is regulated by the source of nitrogen and the time of application, indicating the interrelationships of various management techniques. In addition, each of the system variables also influences the optimum placement of nitrogen fertilizers.

Soil features such as CEC, texture, organic matter content, water-holding capacity, and pH have a major impact on the placement of nitrogen fertilizers for efficient use. These features regulate volatilization loss and nitrogen movement. Smith (1966) identified essentially the same factors and concluded that depth of placement has a major impact on the retention of ammonium sources in soils. If the soil is acid enough and has sufficient buffering capacity, the potential for volatilization losses of ammonium is reduced. Adams and Martin (1984) indicated that pH and environmental factors will affect the chemical reaction of ammonia with soil-water and clay surfaces. Increasing pH increases ammonia loss by shifting the reaction towards ammonia gas, as does drying the soil. Reaction of ammonium with the clay surfaces by cation exchange shifts the reaction in such a manner as to retain ammonium in the soil. Nitrate-N is not subject to volatilization loss. The relative rates of losses as affected by each of the variables that have been discussed are depicted graphically by Rolston (1978), see Fig. 7-1. It is evident from this figure that differences in placement and pH have the greatest impact on volatilization losses of ammonia from soils.

There are basically two types of fertilizer placement: broadcast application on the surface, or in bands below the surface. Such terminology as top dress, side dress, broadcast, list into beds, starter applications, chisel injection, banding, or deep placement are simply variations of these two techniques resulting from different methods of mechanical distribution, depth of placement, and timing in relation to planting of the seed and development of the plant.

One system variable that has a major impact on placement of nitrogen is the type of crop grown. This is exhibited in a number of ways. Placement of fertiliz-

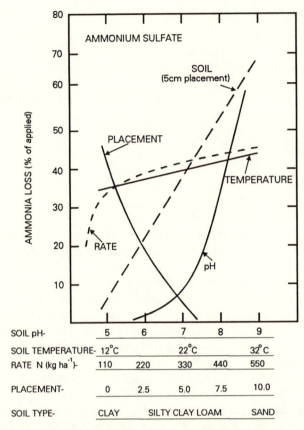

Figure 7-1 Approximate relationships of ammonia loss to soil pH, soil temperature, ammonium application rate, depth of placement, and soil type for ammonium
sulfate. The relationships are for ammonia loss from a clay soil (excluding soil type
curve), surface applications (excluding placement curve), 550 kg/ha ammonium-N
application rate (excluding ammonium rate curve), 30°C (excluding temperature
curve), and pH 7.6 (excluding pH curve). (Rolston, 1978.)

ers with a high salt index in proximity to the seed or root system impacts plant
growth. Salt indexes for a variety of fertilizer materials have been indicated in
Table 5-12. In general, plants that have higher salt tolerances are able to withstand larger amount of fertilizers with high salt indexes placed in their proximity.

Another aspect of crop tolerance is the specific ion effect on plant development. Even a material with a low salt index may possibly have an adverse effect
on seed germination and plant development resulting from specific ion toxicity.
The placement of ammonium fertilizers, as shown in Fig. 7-2, is done to minimize the effects described.

Salt and specific ion toxicities also influence the use of foliar applications of
various materials. In Table 7-1 is shown the tolerance of the plant foliage to
urea sprays, indicating, in general, that plants with higher salt tolerances can
withstand higher application rates of urea sprays on their foliage. Foliar appli-

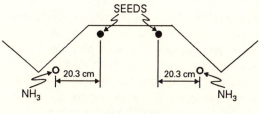

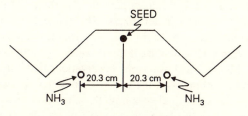

Figure 7-2 Placement of preplant ammonia fertilizer to avoid toxicity to young plants. Place fertilizer band at, or just below, the level of the furrow bottom. (Sailsbery and Hills, 1976.)

Table 7-1.   Tolerance of plant foliage to urea sprays in pounds per 100 gallons of water.

Crop	Tolerance
Cucumber	3-5
Tobacco	3-10
Apple, bean, grape, lettuce, pepper, raspberry, strawberry, sweet corn, tomato	4-6
Banana, cacao, citrus	5-10
Plum	5-15
Cherry, corn, peach	5-20
Cabbage	6-12
Sugarcane	10-20
Alfalfa, carrot, celery, onion, potato, sugarbeet	10-20
Cotton, pineapple	20-50
Bromegrass, wheat	20-800
Hops	40-50

Source: Adapted from Wittwer et al., 1963.   Copyright © 1963 by Soil Science Society of America. Reprinted by permission.

cation as a form of placement may be effective as a means of fertilization, as shown by Soltanpour (1969). He found the relative marketable yield of potatoes receiving approximately 140 kg/ha of N as all-foliar application (12 applications), all-soil-preplant, and 50 percent preplant–50 percent foliar applied (12 applications) resulted in relative yields of 76, 83, and 100 percent, respectively. This indicates that it is technically possible to make several applications of low-concentration nitrogen solutions to supply a major portion of the crop nitrogen without resulting in severe damage due to either the salinity or specific ion effects. However, it is economically impractical because of the cost of such a large number of foliar applications.

Rooting depth and root distribution play an important role in determining the optimal placement of nitrogen fertilizers. The range of rooting depths for a variety of crops is shown in Table 7-2. The rooting depths of young plants would be considerably less, making it more critical for nitrogen fertilizers to be properly placed. Figure 7-3 shows some observations that indicate the impact of rooting depth and distribution on the uptake of nitrogen fertilizer from different placements. Figure 7-3 also indicates how these relationships influence the distribution of the nitrogen and carryover in the soil profile. At very low nitrogen fertilizer rates, placement had little impact on the nitrogen carryover. The further the placement is away from the plant row at excessive rates of fertilization, the higher the concentration of nitrogen at the outer reaches of the root system. Even if nitrogen is placed in the most active portion of the root system, the plant is incapable of extracting all the nitrate nitrogen out of that zone if the rate of application is excessive.

The method of irrigation and pattern of water application interact with soil infiltration characteristics to produce a major impact on placement. Water movement patterns and the resultant nitrate distribution are indicated in Figs. 4-3 to 4-6. Because of spatial variabilities that exist in soils within a management unit, irrigation method and placement ultimately determine the efficiency of use and magnitude of nitrate loss.

The source of nitrogen also influences the transformation it undergoes, the soil environment, and subsequently, the placement. As is evident in Fig. 5-1,

Table 7-2.   Expected depth of rooting for various types of crops.

Crops	Range of rooting depth, cm
Irrigated pasture, onions, strawberries	30 - 60
Cucumbers	45 - 75
Beans, cereal grains, pepper	60 - 120
Alfalfa, corn, sugar beets, tomatoes	90 - 150
Grapes	90 - 180
Sudan, watermelons	120 - 180
Deciduous fruits and nuts (except walnuts)	150 - 210
Walnuts	180 - 300

Source: Anonymous, 1977.

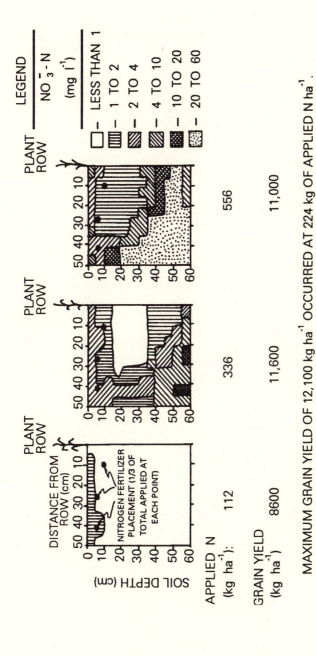

Figure 7-3 Distribution of nitrate-N in soil beneath corn (*Zea mays*) at harvest (A.B. Carlton, University of California, San Joaquin Valley Research and Extension Center, personal communication, 1973.)

there are a large number of simultaneous reactions occurring in soils that determine the amount and form of available nitrogen for the plant. The overall tendencies of these transformations is to produce nitrate-N, which is very mobile in the soil environment. As already shown, the rate of nitrification is sufficiently rapid under most growing conditions to insure quick conversion of ammonia to nitrate. This may contribute to a higher potential for nitrogen loss from the root zone. One mechanism that has been utilized to reduce the rate of nitrification has been the placement of ammonium fertilizers in bands below the soil surface. The relatively high concentration of ammonium in a small soil volume increases the soil pH and reduces the rate of nitrification. Table 7-3 shows the influence of the placement of urea in fine sandy soil on the rate of hydrolysis to ammonium. Ammonium moved in all directions from the point of placement as a function of a concentration gradient. Because of the elevated pH in the zone of highest concentration, the formation of nitrate was inhibited so that only low soil nitrate levels were found. Figure 7-4 shows the distribution of ammonium and nitrate in a cross-section of four different soils with different initial soil pH values and utilizing four nitrogen sources. After 14 days, even with a band placement, the transformation of ammonium to nitrate was nearly complete. This indicates that while band application can reduce the rate of nitrification, it does not prevent it. Nitrification inhibitors placed in bands with ammonium fertilizers have a greater inhibitory effect on the nitrification rate, but even this does not prevent the process entirely.

The effect of placement on plant uptake differs with nitrogen source (Table 7-4). Much higher recoveries and crop uptakes resulted as a function of banding

Table 7-3.   Field tests of urea[*] mobility and transformations when band placed at 7.5 cm in a Leon fine sand[b].

Soil depth	2 Days after placement			6 Days after placement			
	Urea-N	NH_4-N	pH	NH_4-N	NO_2-N	NO_3-N	pH
cm	-------------ppm---------------			--------------------ppm-------------------			
0-1.3	--	2	6.6	1	--	13	5.8
1.3-3.2	--	0	6.8	151	0.1	4	7.8
3.2-5.1	--	59	8.0	291	0.1	3	8.3
5.1-6.4	241	360	9.1	332	--	2	8.6
6.4-7.0	363	469	9.1	363	0.2	2	8.6
7.0-8.3	135	493	9.1	364	0.2	2	8.7
8.3-8.9	34	399	9.1	337	0.2	2	8.8
8.9-10.2	--	199	8.9	302	0.5	2	8.8
10.2-12.1	--	8	7.0	166	1.1	4	8.4
12.1-15.2	--	0	6.8	--	--	4	6.4

Source: Adapted from Volk, 1965.

[*] 1.95 g of urea per 187 cm[2] was the actual application rate which was estimated to be comparable to 84 kg of N per ha for two bands per row on 87-cm centers.

[b] Soil moisture was approximately 75% of field capacity at the point of placement.

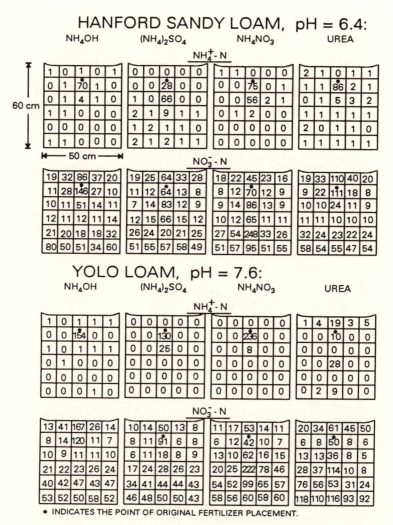

Figure 7-4A Distribution of ammonium-N and nitrate-N in a vertical cross-section of soil when different fertilizers are band-placed in Hanford sandy loam (coarse-loamy, mixed, nonacid, thermic Xerorthents) and Yolo loam (fine-silty, mixed, non-acid, thermic type Xerorthents) soils. (Tyler et al., 1958. Copyright © 1958 by the American Society of Agronomy. Reprinted by permission.)

ammonium sulfate, ammonium nitrate, or aqua-ammonia below the surface as compared to surface application of these materials. Similar effects of band placement have been observed for rice as shown in Table 7-5. Where comparisons between broadcast and band-applied nitrogen were made, there was a substantial yield increase for the band-applied nitrogen. This indicates a higher use efficiency, presumably as a result of minimizing ammonia volatilization loss. A similar observation was made by Soltanpour (1969) for potatoes. He found that apparent recovery of applied nitrogen was reduced from approximately 75 per-

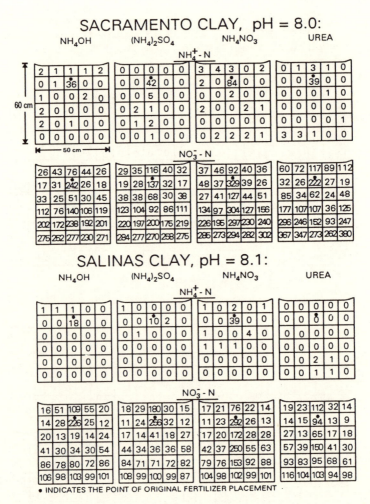

Figure 7-4B Distribution of ammonium-N and nitrate-N in a vertical cross-section of soil when different fertilizers are band-placed in Sacramento clay (very fine, montmorillonitic, thermic vertic Haplaquolls) and Salinas clay (fine-loamy, mixed, thermic pachic Haploxerolls) soils. (Tyler et al., 1958. Copyright © 1958 by the American Society of Agronomy. Reprinted by permission.)

cent when ammonium sulfate fertilizer was banded compared with approximately 14 percent when it was broadcast.

Climatic effects have also been demonstrated to have an impact on placement. Temperature influences the rates of chemical reactions as well as rates of biological transformations. At lower temperatures, the magnitude of volatilization losses is generally reduced. Temperature has an even greater influence on biological transformation of nitrogen, such as the hydrolysis of urea and nitrification of ammonium. This in turn plays a significant role in influencing the placement of nitrogen fertilizers. For example, urea may be placed on the surface of soils when temperatures are below 0°C without concern for volatilization

Table 7-4.   Average percent of crop uptake, residual N in soil, and overall recovery of fertilizer N for three soils with surface and band applied fertilizer.

Placement	Ammonium sulfate[a]	Ammonium nitrate[a]	Ammonium nitrate[b]	Potassium nitrate	Aqua ammonia	Urea
			----Percent----			
Surface:[c,d]						
Crop uptake[e]	48.3	52.3	65.4	66.3	35.5	58.2
Residual	23.2	30.5	27.3	27.3	21.3	35.0
Recovery	71.5	82.8	92.7	93.6	56.8	93.2
Banded:[f]						
Crop uptake[g]	70.0	72.6	---[h]	---	66.4	---
Residual	22.1	26.3	---	---	31.2	---
Recovery	92.1	98.9	---	---	97.6	---

Source: Adapted from Broadbent and Nakshima, 1968. Copyright © 1968 by Soil Science Society of America. Reprinted by permission.

[a] Ammonium nitrate tagged on the ammonium ion.
[b] Ammonium nitrate tagged on the nitrate ion.
[c] Fertilizers were dissolved in water and uniformly applied to the soil surface in solution and then irrigated.
[d] [15]N tagged fertilizer materials were used in the greenhouse.
[e] Crop uptake is based on a sequence of four crops over a period of 314 days, including two cuttings of sudan, a tomato and a corn planting.
[f] [15]N tagged fertilizer materials were used in the greenhouse.
[g] Crop uptake is based on two cuttings of sudan over a period of 75 days.
[h] Dash line indicates no data.

Table 7-5.   Effect of method of nitrogen application on rice yield index for various soils.

		SOILS				
Fertilizer treatment	Rate of applied N	Stockton clay	Genevera clay	Playa silty clay	Willows clay	Rocklin clay
	kg/ha	--------------------------Rice yield index[a]--------------------------				
No applied nitrogen	(ON)	100	100	100	100	100
Broadcast in water	(34N)	108				
Broadcast dry seed bed	(34N)	119				
Drilled	(34N)	137				
Broadcast dry seed bed	(45N)		136	145	125	172
Drilled	(45N)		163	165	147	204
Broadcast dry seed bed	(67N)		158	148	141	235
Drilled	(67N)		225	167	178	248

Source: Adapted from Mikklesen and Finfrock, 1957. Copyright © 1957 by American Society of Agronomy. Reprinted by permission.

[a] Rice yield index is the percent yield increase over the no applied N.

164

loss since the enzyme necessary for the hydrolysis cannot function below 0°C. Similarly, nitrification does not proceed at soil temperatures below 0°C, which impacts both the selection of nitrogen source, and timing and placement of the applications.

The frequency and the amount of rainfall influences nitrogen fertilizer by virtue of its impact on the rate of movement through the soil profile and the potential for either denitrification or leaching losses. In areas where precipitation potential is high and frequent there would be a greater potential for denitrification and leaching loss. In either case, band application of an ammonium form of fertilizer is the method of placement to mitigate such effects.

It is important to understand how different irrigation methods and mechanical manipulations influence the distribution of nitrogen in the soil profile in order to ascertain the appropriate placement to enhance nutrient uptake by plants. Figure 7-5 shows schematically what happens when different nitrogen sources are subjected to a variety of placements, incorporation, and irrigation methods. In Fig. 7-5(A) fertilizer nitrogen is broadcast on the soil surface. Regardless of the nitrogen source used and the magnitude of the volatilization loss, the remaining nitrogen would be distributed in beds approximately in the manner indicated.

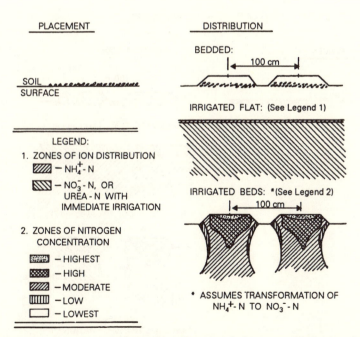

A. SURFACE BROADCAST WITH NO INCORPORATION:

Figure 7-5A Schematic showing the influence of placement of nitrogen fertilizers on nitrogen distribution in soil for surface broadcast without incorporation into the soil.

This presumes that the beds were mechanically formed after the surface application and prior to any precipitation or irrigation.

If the nitrogen fertilizer had been broadcast on the surface and the soil, then irrigated flat, assuming that fertilizer contained both ammonium and nitrate nitrogen, the distribution of ammonium and nitrate in the soil profile would be similar to that indicated. Most of the ammonium would be adsorbed relatively near the surface and the nitrate would move into the soil profile and be distributed behind the wetting front. This particular part of the schematic is intended to show only relative concentration. Zones of relative nitrogen concentration are shown for irrigated beds. In Fig. 7-5(B) nitrogen fertilizer is assumed to be disked after broadcasting, resulting in the placement of the nitrogen throughout the zone a few centimeters below the soil surface. A bedding operation distributes the nitrogen fertilizer in a uniform manner in the top of the bed. This type of placement results in a slightly different distribution of nitrate as shown in Fig. 7-5(A) after an irrigation.

Figure 7-5(C) shows the effect of different band placements after irrigation. Ammonium moves very little from point of placement and nitrate distribution broadens as it moved down through the profile. In the case of irrigated beds higher concentrations of nitrate occur in the middle of the beds, basically in the zone where the bands have been placed. In most cases, this zone extends to the surface of the bed, and, as a result, across the bed. This placement and nitrate

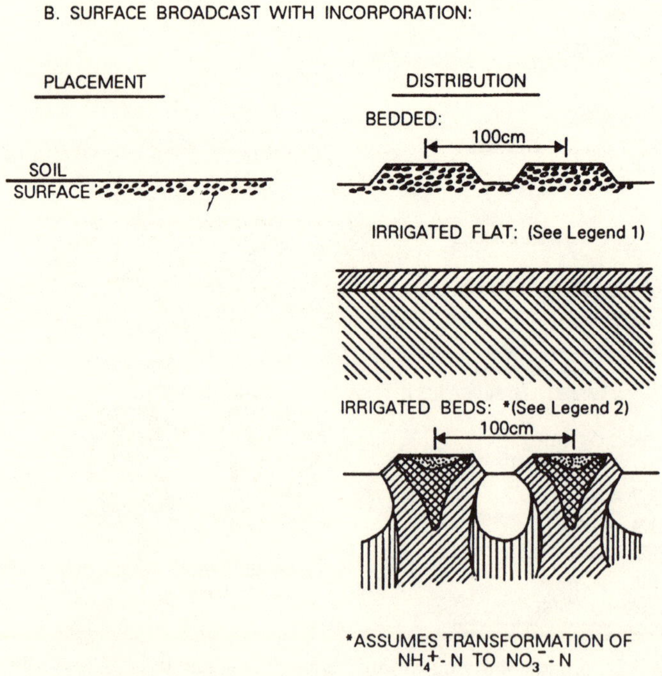

Figure 7-5B Schematic showing the influence of placement of nitrogen fertilizers on nitrogen distribution in soil for surface broadcast with incorporation into the soil.

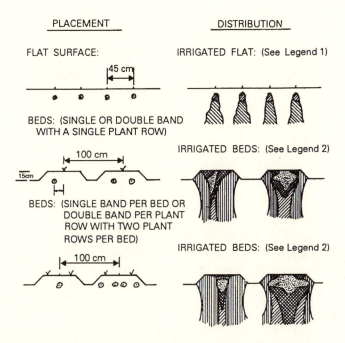

C: INJECTED IN BANDS:

PLACEMENT   DISTRIBUTION

FLAT SURFACE: IRRIGATED FLAT: (See Legend 1)

45 cm

BEDS: (SINGLE OR DOUBLE BAND
WITH A SINGLE PLANT ROW)

100 cm   IRRIGATED BEDS: (See Legend 2)

15cm

BEDS: (SINGLE BAND PER BED OR
DOUBLE BAND PER PLANT
ROW WITH TWO PLANT
ROWS PER BED)   IRRIGATED BEDS: (See Legend 2)

100 cm

Figure 7-5C Schematic showing the influence of placement of nitrogen fertilizers on nitrogen distribution in soil for injected bands of fertilizer in the soil.

movement with furrow irrigation regulates the movement of the nitrogen such that most of the nitrogen stays in the bed.

By utilizing a variety of placement techniques it is possible to manipulate the distribution of nitrogen in a soil profile in order to influence the rate of nitrogen transformations, denitrification, and nitrification, and minimize volatilization or leaching losses and enhance the possibility of greater plant uptake. The overall impact is increased nitrogen-use efficiencies and a decrease in the nitrate pollution potential of an agriculture system.

8

Equipment

The type of machinery used for fertilizer application is seldom considered as a management variable. However, the selection of a piece of equipment for fertilizer application plays a definite role in the determination of the fertilizer source that can be used. Since the source of fertilizer nitrogen plays such an important role in the placement and timing of nitrogen fertilizers, and because of the different physical states (solid, liquid, and gas) that occur for the major nitrogen fertilizers materials, the selection of the proper piece of equipment or the selection of the proper source for a given piece of equipment makes the type of equipment available for fertilizer application a factor the grower must consider in the management of nitrogen. There are a large number of different fertilizer applicators designed to do one of two things—either apply the fertilizer to the surface of the soil or inject the fertilizer below the surface. There are a wide variety of metering and spreading devices used for both techniques in order to apply the material in as uniform a manner as possible.

Each of the system variables has some influence on the type of the equipment that might be selected for a nitrogen fertilizer application. Soil structure and texture, to the extent that they govern the ease of tillage and the type of implements that might be used to incorporate the fertilizer, influence equipment selection. For example, in a medium or fine textured soil with a massive structure the energy required for injection of fertilizer in bands below the surface, and the wear on the implements, might dictate that materials would be surface broadcast and incorporated by disking, plowing, or irrigating.

Soil-water content also influences equipment selection. If the soil is saturated or nearly saturated, then it may be necessary to broadcast the fertilizer by aerial application. At lower water contents (soils at field capacity or slightly drier), the possibility exists for utilization of high flotation equipment for broadcast application. Even though it is mechanically possible to inject gaseous ammonia in very wet soils, it is generally undesirable because of the inability of the soil to fill in behind the injector shanks. Soils with water contents in the range of 50 to 75 percent of available water-holding capacity have few restrictions as to type of equipment that may be used. As the soil approaches very low water contents, injection of anhydrous ammonia is again undesirable because the soil fractures as the applicator shank moves through the soil, which results in a high potential for volatilization losses.

The type of crop and the rate of the crop development also place some limitations on the type of equipment that can be used for fertilizer application. Some crops are closely planted, preventing entry into the field. Even with crops not closely spaced, as the plant matures, it either grows too tall or the canopy closes over between the rows, which makes it undesirable to enter the field with the fertilizer applicator equipment because of its potential for damaging the crop. With closely spaced plants or crops with complete canopy cover, post-planting fertilizer applications have to be accomplished by aerial application or application in the irrigation water.

Irrigation method also plays a role in determining the type of equipment selected for fertilizer application since the method governs the frequency of irrigation, and as a result, the water content of the soil, which in turn determines the type of equipment that must be used to apply fertilizer. The type of equipment used to distribute water for a particular irrigation system, such as sprinkler or drip irrigation, will also dictate whether or not it is possible to enter the field after the irrigation system is in place. A variety of equipment exists for applying nitrogen in the irrigation water. Some of the advantages and disadvantages of doing this will be discussed under other management variables.

Consequently, the type of equipment that may be used is limited by the source of nitrogen that is available; conversely, the equipment on hand for fertilizer application dictates the source of nitrogen that may be used. Fortunately, in many parts of the country fertilizer dealers may provide, as a service, fertilizer applicators for the nitrogen sources they handle if the grower would not normally have that type of equipment available. In isolated areas and where traditional use of a particular source has mitigated against the use of other sources, the equipment may not be readily available for the application of an alternate source.

Various methods for ground application of fertilizer materials are shown in Table 8-1 with an indication of the time required to fertilize different land areas and a relative cost index for different methods of application. Even though the values in Table 8-1 are based on 1962 costs, it is assumed that the time required for each practice is the same, and that the costs for each practice have increased in approximately the same relative manner; therefore, the cost index should be a reasonable estimate of the relative cost of each of the fertilizer practices. This assumption may not be entirely true, since there have been improvements in loading equipment and some changes in the relative costs of different nitrogen sources. The data indicate economies of scale, since it is normally assumed that the larger the area fertilized the lower the cost per unit area for the fertilizer application.

The data in Table 8-1 do not show a comparative figure for the cost associated with aerial application of fertilizer. The cost may vary, but is usually in the range of $7.50 to $15.00 (1990 dollars) per hectare plus the cost of the material. There is an upper limit on the amount of material that can be applied in this manner. Consequently, aerial application is generally used as a top dressing with fertilizer rates seldom exceeding 50 kg/ha of N. The material used can be either

Table 8-1.   Approximate time required and relative costs[a] for fertilizer applications using different methods and materials calculated for land area fertilized per year.

| Method of Application and Material Used | Area fertilized - hectares | | | | | | | | | |
| | 10 | | 30 | | 90 | | 270 | | 810 | |
	Time in hours	Cost index	Time in hours	Cost index	Time in hours	Cost index	Time in hours	Cost index	Time in hours	Cost index
Broadcast with a drop flow spreader:										
675 kg of material/ha	8.1	8.38	24.2	4.16	71.6	2.41	215.4	1.88		
Broadcast with an end gate rotor:										
675 kg of material/ha	5.8	6.22	17.5	3.06	49.1	1.76	155.3	1.37	--	--[b]
Surface liquid spray: 67 kg of N/ha as										
Urea-ammonium Nitrate	2.8	11.59	8.4	4.69	24.2	1.85	72.8	1.00[c]	--	--
Preplant injection: 134 kg of N/ha as										
(a) anhydrous ammonia	--	--	--	--	48.2	8.56	145.3	4.69	434.8	3.21
(b) aqua ammonia	--	--	--	--	38.8	8.49	115.2	4.37	344.8	2.76
Sidedress application: 112 kg of N/ha as										
(a) anhydrous ammonia	--	--	19.8	11.94	58.4	5.63	175.1	3.19	525.4	2.51
(b) aqua ammonia	--	--	18.0	10.50	69.5	5.37	208.0	3.37	624.0	2.82
(c) ammonium sulfate	--	--	4.3	6.32	130.8	5.15	391.6	4.75	1174.7	4.63

Source: Adapted from Schade et al., 1962.

[a], [c] The relative cost is a calculated cost index where the lowest cost for application of $0.68 was set equal to 1.00. The cost figures were based on 1962 values, but it is assumed that all costs will increase the same for each input; therefore, the cost index is the indication of the relative cost of fertilization when the same methods of application are used.

[b] The dash lines indicate no data were calculated for these combinations.

a solid or a liquid form of nitrogen fertilizer. One of the problems associated with aerial broadcast application is the difficulty in achieving a relatively uniform application. Examples of typical distribution for single and multiple runs are given in Fig. 8-1. As can be seen, even with a relatively symmetrical distribution pattern for a single run, when multiple runs are made using the same material and same applicator wide variations in the amount of material applied can be observed.

This difficulty is not only associated with aerial broadcast application but with other broadcast applicators as well. Using the example shown in Fig. 8-1, the difference between the lowest and highest rate of fertilizer application in the area fertilized is approximately 40 kg/ha of N. The impact of this variability in broadcast application can be seen in the curves shown in Fig. 8-2. If it is assumed that the rate of application desired is just adequate for maximum yield, then the curve shows that the amount of nitrogen in excess of that required for maximum yield produces no yield increase.

Conversely, where the amount of nitrogen applied is not adequate, then a yield loss is experienced. However, when calculating the amount of nitrogen applied over the total area, there may have been adequate nitrogen to obtain

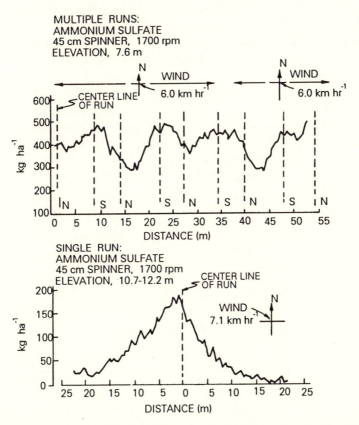

Figure 8-1 Example of distribution of fertilizer broadcast by aircraft. (Brazelton et al., 1970.)

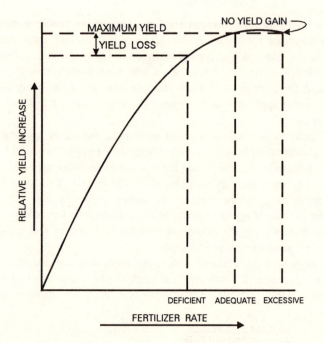

Figure 8-2 Theoretical effect of uneven nitrogen application on yield. (Jensen and Pesek, 1962. Copyright © 1962 by the Soil Science Society of America. Reprinted by permission.)

maximum yield had the material been uniformly applied. With such nonuniformity of fertilizer application, the only way to avoid yield loss is to apply sufficient quantities that the lowest amount applied is still adequate. The result is that a portion of the field is overfertilized, resulting in an increase in cost of production and an increase in the nitrogen pollution potential from that portion of the field that is overfertilized.

9

Rate of Nitrogen Application

Many of the soil physical, chemical, and biological characteristics influence the amount of nitrogen that may be required for plant growth. These interact with the other system and management variables to determine the rate of fertilizer required in order to achieve maximum yield. Even though these soil characteristics contribute to a high degree of variability in soil nitrate levels, soil tests for nitrate have been surprisingly effective guides for estimating the rate of fertilization. For example, in Figs. 9-1 and 9-2 are shown correlations between the soil nitrate-nitrogen concentration and the yield of corn and sorghum, respectively, based on samples obtained from the 0 to 120-cm soil depth. In Table 9-1 are also shown data indicating the relationship between the initial soil nitrogen level and crop yield. These data demonstrate the utility of soil test information for predicting fertilizer application rates. Nevertheless, it must be recognized that other characteristics of the agricultural production system have an impact on the efficient utilization of soil and fertilizer nitrogen and also must be considered in predicting the fertilizer requirements by using these tests.

In the Pacific Northwest, a system developed by James (1971) has been used extensively for estimating the nitrogen fertilizer requirements for a variety of

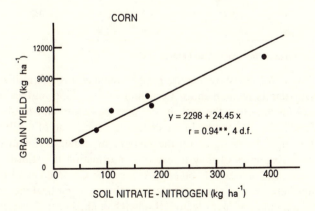

Figure 9-1 Regression of corn (*Zea mays*) yields on soil nitrate nitrogen in 0–120 cm soil depth. (Nelson et al., 1965.)

173

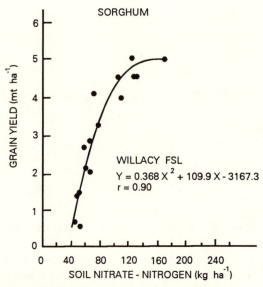

Figure 9-2 Relationship between grain sorghum (*Sorghum vulgare*) yield and nitrate-N in the 0–120 cm Willacy fine sandy loam (fine-loamy, mixed, hypothermic udic Argiustols) soil profile. (Hipp and Gerard, 1971. Copyright © 1971 by the American Society of Agronomy. Reprinted by permission.)

Table 9-1.   Influence of soil nitrogen levels on barley yields response to nitrogen fertilization.

		Soil sample depths in cm					
		Before planting			After harvest		
Nitrogen application rate	Yield	0 - 30	30-60	60-90	0-30	30-60	60-90
kg/ha	kg/ha	------------------------ppm NO₃-N------------------------					
0	4,663	28.6	12.4	3.9	7.5	5.7	2.9
168	5,974	56.6	23.3	5.6	4.3	3.0	2.0
336	6,198	92.4	41.2	13.8	8.4	5.5	3.5

Source: Adapted from Osterli and Meyer, 1976.

crops. In Table 9-2 are shown some sample calculations using this procedure. The fertilizer rate is recommended from this value, taking into account adjustments based on the crop, the season, the previous crop, and fertilizer history. Hills (1976) developed a method for calculating the amount of fertilizer nitrogen required for sugar beets based on the amount of nitrate-N found in a 90-cm depth of soil. Examples of how to use this information in estimating fertilizer rates are shown in Table 9-3. In other states, such as Colorado and Idaho, similar relationships between soil nitrate-nitrogen levels and sugar beet yields have been reported by a number of investigators (Ludwick et al., 1977; Carter et al., 1974). The purpose of reporting so many examples where relatively good correlations have been found between soil nitrate-nitrogen and crop yield is to indicate that

Table 9-2.   Sample calculations of kg of nitrate nitrogen per hectare based on soil
               test results.

Example 1: Where soil depths are in 30 cm increments.

Soil depth	Soil test for NO_3-N	x	Conversion factor	=	Estimated kg of NO_3-N/ha
-----cm-----	---ppm---				
0 - 30	8		4.5		36
30 - 60	12		4.5		54
60 - 90	4		4.5		18
Total					108

Example 2: Where deepest soil layer sampled is less than 30 cm.

Soil depth	Soil test for NO_3-N	x	Conversion factor	=	Estimated kg of NO_3-N/ha
-----cm-----	----ppm----				
0 - 30	8		4.5		36
30 - 60	12		4.5		54
60 - 80	4		3.0[a]		12
Total					102

Source: Adapted from Dow et al., 1969, as cited by Gardner, 1971.

[a] Conversion factor equals: $\dfrac{\text{Depth of Soil Increment}}{30} \times 4.5$

For example, $\dfrac{80 - 60}{30} \times 4.5 = \dfrac{20}{30} \times 4.5 = 3.0$

even while extreme variability is known to exist in soils for nitrate concentration these testing methods have been used with moderate success as guides for predicting fertilizer rates.

Several characteristics associated with crops influence the rate of nitrogen fertilizer required to obtain maximum yield. One such characteristic is the amount of nitrogen removed by the harvested portion of the crop. Because of growing season length, type of plant, and portion with the plant harvested, there can be a wide variation in the total amount of nitrogen removed from the field. This plays a role in the determination of nitrogen fertilizer efficiency that may be obtained from the nitrogen fertilizer applied. The relationship between the amount of nitrogen in the harvested crop and the rate of nitrogen required to obtain that yield determines the amount of residual nitrogen that may remain in the soil after harvest. As shown in Fig. 9-3, there is a significant decrease in the yield of sucrose from sugar beets once the rate of nitrogen fertilizer required to achieve maximum yield is exceeded. In Table 9-4 are shown several crops with their sensitivity to excess applied nitrogen. The possibility of yield decrease for excessive nitrogen is an added incentive to manage nitrogen so as to avoid the addition of excessive amounts. Tolerant crops are able to withstand

Table 9-3.   Nitrogen fertilizer rates estimated from soil nitrate nitrogen (NO₃-N) and expected sugar beet root yield*.

Soil NO₃-N per ha-90 cm	Calculated root yield	kg/ha of Fertilizer N[b]	
		Projected root yield 70 mt/ha	Projected root yield 80 mt/ha
kg	--mt/ha--	---kg/ha---	---kg/ha---
0	46.0	190	270
50	50.4	160	240
100	54.8	120	200
150	59.2	90	170
200	63.6	50	130
250	68.0	20	100
200	72.4	0	50

Source: Adapted from Sailsbery and Hills, 1977.

* Calculations were based on the formula Root yield (T/A) = 20.5 + 0.044x where x = pounds of NO₃-N per acre three feet of soil. The formula was transformed into metric units so Root yield (mt/ha) = 46 + 0.088x, where x = kilograms of NO₃-N in a hectare of soil 90 cm deep.

[b] To calculated fertilizer requirements approximately 8 kg of N is required to produce 1 mt/ha of sugarbeet root. The difference between yield based on soil NO₃-N levels and projected yield was multiplied by 8.

large excesses of nitrogen before any measurable yield decrement can be observed.

The number of plants per unit area has also been shown to influence the response to applied nitrogen. For example, in Fig. 9-4 as the number of plants per hectare increased, the yield of corn responded differently to three nitrogen levels. The highest yield was obtained with a high nitrogen level and a relatively large number of plants per acre. Unpublished data by Rauschkolb and Dennis (1969) in Arizona showed that high seeding rates of wheat in unfertilized plots out-yielded low seeding rates. At adequate applied nitrogen level for maximum yield, there were no differences between a low and high seeding rate.

Another predictive tool that is available for determining the rate of nitrogen fertilizer required for maximum yield is the tissue test. Most investigators have utilized tissue tests to indicate either a deficiency or sufficiency of nitrogen for plant growth. Few investigators have utilized the tissue test to determine the rate of nitrogen that would be required to obtain maximum yield once a deficiency has been detected. Fertilizer recommendations have been developed based on tissue test levels for nitrate-N at different stages of plant development. Examples of these approaches are shown in Tables 9-5 and 9-6 for lettuce and sugar beets. In addition to making fertilizer recommendations based on the nitrogen content in a particular tissue at a certain stage of development, this approach has been taken one additional step by combining that information with the initial soil test. This approach is shown in Table 9-5 and represents a refinement in

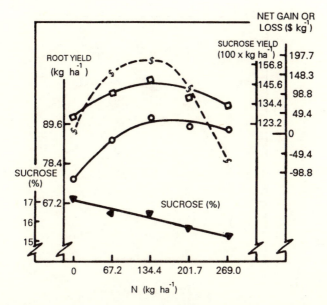

Figure 9-3 Profit is greatly affected by nitrogen fertilization. On this field 134 kg/ha of nitrogen gave maximum net gain; more nitrogen decreased profits. Net gain or loss is gross value less the gross value of unfertilized beets, the cost of fertilizer and its application, and harvest and haul charges for the difference in production between fertilized and unfertilized beets. (Data uased: $0.44 per kg nitrogen; $0.61 per ha for application of nitrogen; $3.30 per metric ton, harvest and haul charge; $0.06 per kg, processor's "net selling price".)

Table 9-4. Relative sensitivity of crop yield and quality to excess nitrogen.

	Sensitive crops[*]		Tolerant crops[*]	
Increasing	Sugarbeets	Citrus	Corn	Lettuce
tolerance to	Onions	Avocados	Sorghum	Spinach
excess N	Potatoes	Grapes	Millet	Broccoli
	Tomatoes	Apricots	Forage grasses	Walnuts
	Wheat	Strawberries	Sweet Corn	Olives
	Barley	Peaches		Almonds
	Rice	Nectarines		Celery
	Cotton	Pears		Asparagus
v				

Source: Adapted from Krantz and Miller, 1968.

[*] Crops are listed as those least tolerant of excesses first and those most tolerant of large excesses last.

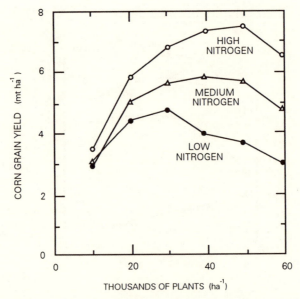

Figure 9-4 The yield of corn (*Zea mays*) at different stand densities and different levels of nitrogen. (Lang et al., 1956 as shown in Mitchell, 1970. Copyright © 1970 by Iowa State University Press. Reprinted by permission.)

Table 9-5. Lettuce nitrogen fertilizer recommendations from emergence to heading stages of plant development based on nitrate-nitrogen content of midribs.

Range of nitrate-N in midrib, ppm	kg/ha of N to be applied up to heading[a]
More than 10,000	none
6,000 to 10,000	30 to 70
3,000 to 6,000	45 to 90
Less than 3,000	70 to 115

Source: Adapted from Openshaw et al., 1973.

[a] Nitrogen fertilizer applications are not recommended when weekly average air temperature is below 13° Celsius (55°F).

combining soil and tissue tests for predicting nitrogen fertilizer application rates. This more closely determines the amount of fertilizer to apply, leading to an increased efficiency of fertilizer and soil nitrogen utilization and with a subsequent reduction in carryovers that reduces nitrogen pollution potential.

In Table 9-7 are shown states having statewide publications for the purpose of recommending nitrogen fertilizer rates. Some utilize soil tests, tissue tests, empirically derived field data, or some combination of these three approaches. The combination of soil and tissue tests corroborated by field investigations is the best approach. The next best technique is the use of tissue tests. However, this has limitations since there may not be adequate plant material available to

Table 9-6.   Fertilizer nitrogen recommended for sugar beets at thinning based on petiole and soil analysis for nitrate-nitrogen.

Range of petiole NO₃-N at thinning ppm	Initial NO₃-N level in soil sample, ppm			
	Greater than 100	20-100	10-20	Less than 10
	--------------------------kg/ha of applied N--------------------------			
Greater than 13,000	0	0	0	0
5,000 to 13,000	0	0	0	0
4,000 to 5,000	*	0 to 25	0 to 25	0 to 45
3,000 to 4,000	*	30 to 45	45 to 70	70 to 90
2,000 to 3,000	*	45 to 70	70 to 90	90 to 135
Less than 2,000	*		90 to 135	135 to 180

Source: Adapted from Openshaw, 1972.

* It is unlikely this condition will occur. If petiole levels are low with high soil test values, additional tests would be indicated.

Table 9-7.   Western states having statewide publications of recommended rates of nitrogen fertilization of crops based on soil testing, tissue testing and/or empirical field data.

State	Recommendation based on:		
	Soil tests	Tissue tests	Field data
Arizona	yes	yes	yes
California	no	no	no*
Colorado	yes	no	yes
Idaho	yes	no	yes
Kansas	yes	no	yes
Montana	yes	no	yes
Nebraska	yes	no	yes
Nevada	no	no	yes
New Mexico	no	no	yes
North Dakota	no	no	yes
Oklahoma	yes	no	yes
Oregon	yes	no	yes
South Dakota	no	no	yes
Texas	no	no	yes
Utah	yes	no	yes
Washington	yes	no	yes
Wyoming	no	no	yes

* Recommendations are published at the local level in county publications.

sample for determining early nitrogen needs of the plants. If no other technique is available, soil tests serve as a guide for making fertilizer recommendations. However, their interpretation must be made with an understanding of the influence of nitrogen transformations, soil, and irrigation method on the form of nitrogen in the soil profile and with consideration of the transient nature of those forms and the spatial variability that normally occurs.

The method of irrigation also has an impact on the efficiency of nitrogen fertilizer utilization, which in turn influences the rate of application. As shown in Fig. 9-5, significant differences in the relative concentrations of inorganic nitrogen below a potato row were found whether it was sprinkler or furrow irrigated. Middleton et al. (1975) found that furrow irrigation caused a more complete removal of applied nitrogen from the root zone than did sprinkler irrigation. This is not surprising considering the amount of leaching that would be expected to occur from the furrow-irrigated plots by virtue of the larger quantity of water applied. The method of irrigation used does have an affect on the amount of water applied and the uniformity of application, which in turn can lead to either differential leaching or enhanced potential for denitrification depending upon the infiltration rate of the soils.

In Fig. 9-6 are shown amounts of added nitrogen that may be subject to leaching losses as a function of fertilizer application rate and resultant plant uptake efficiencies. At low rates of nitrogen application, there is a high efficiency of uptake and only small differences in the amount of nitrogen subject to leaching. As the fertilizer rate increases, efficiency declines. In Fig. 9-7 a similar relationship is shown. Relatively high efficiencies of fertilizer uptake occurred for the first and second increment of nitrogen applied. Maximum yield occurred at 224 kg/ha of applied nitrogen. There was no additional yield increase

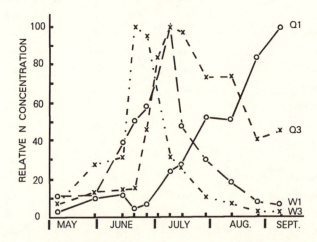

Figure 9-5 Relative concentrations of dissolved inorganic N based on average concentrations in cup samplers placed under potato (*Solanum tuberosum*) rows at depths of 61 and 122 cm with different applications of sprinkler (Q) and furrow (W) irrigation. Maximum peak concentrations were set equal to 100. (Middleton et al., 1975.)

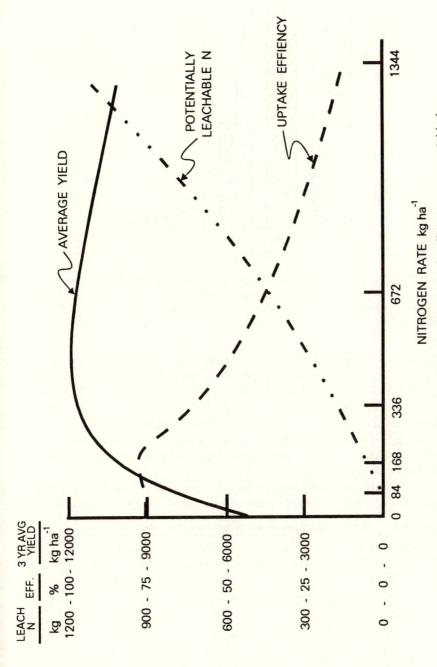

Figure 9-6 The influence of applied nitrogen on uptake efficiency average yield of orchard grass (*Dactylis glomerata*) and residual nitrogen. (Donohue et al., 1973. Copyright © 1973 by the American Society of Agronomy. Reprinted by permission.)

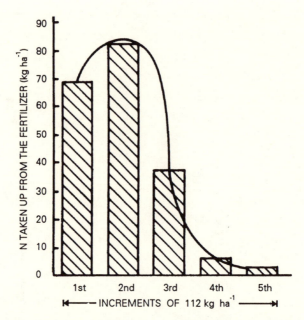

Figure 9-7 Nitrogen taken up by corn (*Zea mays*) from successive increments of added fertilizer N, Kearney Station, California, 1973. (Fried et al., 1976. Copyright © 1976 by the American Society of Agronomy, Inc., the Crop Science Society of America, and the Soil Science Society of America. Reprinted by permission.)

as a result of subsequent increments of fertilizer and the efficiency of plant uptake of each added increment dropped substantially. It is interesting to note that even at extremely high rates of application the plant was still utilizing a portion of the applied nitrogen. As shown in Table 9-8, for each year the maximum soil nitrogen uptake occurred at the fertilizer rate providing maximum yield. These and similar data demonstrate that uptake of the indigenous soil nitrogen is enhanced as the fertilizer rate is increased to supply adequate, but not excessive, nitrogen for maximum yield.

Crop adaptability to lower temperatures determines whether the plant is capable of growing during a period of low temperatures, and this can influence

Table 9-8.   Uptake of soil N by corn as affected by fertilizer rate, Kearney site.

Fertilizer Rate	1973	1974	1975
	----kg/ha----		
0	34	35	40
112	63	36	59
224*	80	37	69
336	66	25	60
448	57	23	43
560	53	18	44

Source: Broadbent and Carlton, 1976.

*Application rate at which maximum yield occurred.

the rate of nitrogen supplied. Some plants have the capability of withstanding temperatures below 0°C, but the growth of the plant is restricted. As discussed in Chapter 6, nitrate uptake by lettuce has been found to be very limited at temperatures below an average weekly temperature of 13°C, but at temperatures above that level lettuce plants will absorb and translocate nitrate. The type of plant that can be grown and its yield potential are affected by climate, which also influences the amount of nitrogen taken up by the plant. Thus, climate is an important factor in determining the rate of nitrogen fertilizer used.

Another feature of soil nitrogen carryover that it is important to examine is the relationship between the soil nitrogen concentration and the soil solution concentration. In Table 9-9 is shown the amount of residual soil nitrogen occurring after corn harvest. There was 80.8 kg of nitrogen distributed in the top 180 cm of soil and this consisted of both residual fertilizer and soil nitrogen. When these amounts of nitrogen are expressed as ppm in air-dry soil, the values range from 0.8 to 6.0 ppm. These values represent an extremely deficient level of soil nitrogen and a level that would result in severe yield reductions if additional fertilizer nitrogen were not applied. However, when these same concentrations in the soil are expressed as ppm nitrogen in the soil solution, only 3 of the 18 depth and water combinations have a nitrate-N concentration below 10 ppm, which is the level established as the criterion for determining nitrogen pollution.

The ability of nitrogen carryover from previous fertilizer applications to provide adequate nitrogen for subsequent crops is demonstrated in Table 9-10. When a relatively high rate of fertilizer was applied in the previous year, the crop still required 112 kg/ha of nitrogen to produce maximum yield in the subsequent crop year. At the lowest application rate for the previous year, the nitro-

Table 9-9.   The distribution of nitrogen in a Hanford soil after harvest of corn and the calculated soil solution concentration at different assumed water contents.

Soil depth in cm	kg of N/ha in the soil increment	ppm N[b] in soil	ppm N in soil solution at different water contents[a]:		
			22%	11%	2.5%
0-30	6.7	1.5	6.7	13.4	26.8
30-60	26.9	6.0	26.0	53.4	107.6
60-90	20.2	4.5	20.2	40.4	80.8
90-120	13.5	3.0	13.5	27.0	54.0
120-150	10.1	2.3	10.1	20.2	40.4
150-180	3.4	0.8	3.4	6.8	13.6

Source: Unpublished data from the University of California NSF-RANN study on Nitrates in Drainage Effluents, 1973.

[a] Soil-water contents after harvest correspond to approximately 22, 11 and 2.5 volume percent, respectively for plots irrigated at 5/3 ET, 1 ET, and 1/3 ET. Calculations of soil solution concentration of nitrogen are based on the assumed contents which are typical for medium to coarse textured soils. ET is the evapotranspiration demand.

[b] These are values for soil nitrogen levels in samples taken after harvest of corn at the Kearney Field Station in 1973. Values are the average of 4 replications for plots fertilized with 224 kg of N/ha which was the rate at which maximum yield occurred.

Table 9-10.   Effect of previous years nitrogen application on response of corn to
               nitrogen fertilization.

Rate applied for current years crop	Yield for current year when rate for previous year was:		
	56 kg of N/ha	112 kg of N/ha	168 kg of N/ha
--------kg/ha--------	----------------------------Yield in kg/ha----------------------------		
0	4,771	5,914	6,989
56	7,526	8,266	9,072
112	9,206	9,811	10,348
168	10,214	10,550	10,819

Source: Adapted from Barber, 1974. Copyright © 1974 by National Fertilizer Solutions
Association. Reprinted by permission.

gen rate required to produce maximum yield was 168 kg/ha. These data demonstrate that the carryover of nitrogen from one season to the next is very limited
under normal growing conditions and where adequate, but not excessive,
amounts of nitrogen fertilizer are applied. Similar results have been observed
by Donohue et al. (1973), who found that there was no difference in the yield
obtained from the plots that had received the nitrogen 3 years previously and
the plots that received no nitrogen at that time. The impact of residual nitrogen
from high application rates in this case was negligible.

It has been demonstrated in several crops by many investigators that where
peak yields occur there is a dramatic increase in the amount of leachable nitrogen resulting from additions of applied nitrogen beyond that required for
maximum yield. Because of the close relationship between the input and output
of nitrogen in a cropping system, the rate of fertilizer application becomes one
of the key factors in regulating the amount of nitrogen carryover in the soil–
plant–water system. Consequently, considerable attention should be given to the
rate of nitrogen fertilizer application for the purpose of adjusting the rate to
provide peak efficiency for any given set of conditions in the agricultural production system.

10

Source of Nitrogen

Because of the physical and chemical characteristics of nitrogen fertilizer materials and the different ways in which each of these materials interacts with the system variables, the source of nitrogen becomes an important factor in nitrogen management. As discussed in Chapter 5, these characteristics influence nitrogen movement in soils, the susceptibility to loss, and the transformation of nitrogen in the soil–plant–water system. Soil physical characteristics influence the rate of nitrogen movement through the soil profile and determine the magnitude of losses as a function of denitrification and leaching. Soil chemical characteristics such as cation exchange capacity (CEC) and pH affect the amount of ammonia volatilization losses that may occur. The amount of organic matter present and the degree of decomposition will influence the amount of nitrogen tie-up (immobilization) or release (mineralization). Each of the factors indicated plays a role in determining the most suitable source of nitrogen to use for any of these conditions.

The interaction of the physical and chemical characteristics for different nitrogen sources as they influence the relative ammonia volatilization loss is summarized in Table 10-1. Relative losses are shown for different soil placement techniques with soil pH above or below 7.0. Ammonium sources are more susceptible to volatilization losses, especially in soils at a pH above 7.0. Urea acts similarly to ammonium because of its rapid enzymatic hydrolysis in soils. However, by virtue of its water solubility and net neutral charge prior to hydrolysis, the application of urea in water, or incorporation with water immediately after a broadcast application, results in relatively low losses because the material is moved well into the soil profile before significant hydrolysis can occur. Nitrate is not susceptible to such losses and where it is present with ammonium or urea the magnitude of ammonia volatilization loss is substantially reduced; see Table 10-2. When any of the sources of nitrogen are placed well below the soil surface, almost irrespective of the soil type, the potential for volatilization loss is very low or nonexistent.

The principle relationships between crop and nitrogen source is the specific ion effect, which may be further categorized into preferential adsorption of nitrogen sources and possible toxic effects. Specific ion toxicities due to nitrates have been alluded to, but there is little evidence in the literature to show the

Table 10-1. Estimates of relative nitrogen losses by ammonia volatilization for different application methods and fertilizer materials.

Nitrogen Source/ Physical State	Nitrogen content	Surface broadcast							Band 10 cm below surface
		Method of incorporation							
		Without water		With water		Mechanical	Apply in irrigation		
		<7[b]	>7[c]	<7	>7		<7	>7	
	---%----								
Anhydrous ammonia, G[a]	82	--[d]	--	--	--	--	L	H	V
Aqua ammonia, L	20	--	--	--	--	--	L	H	V
Ammonium sulfate, S	21	L[e]	H	L	H	L	L	H	V
Ammonium phosphate, S	11	L	H	L	M	L	L	M	V
Ammonium nitrate, S	33.5	V	L	V	L	V	V	V	V
Urea-ammonium nitrate, S	32	V	M	V	V	V	V	V	V
Urea, S	45	M	H	V	V	L	V	V	V
Calcium nitrate, S	15.5	V	V	V	V	V	V	V	V
Potassium nitrate, S	13	V	V	V	V	V	V	V	V

[a] Physical states are represented by G = gaseous, L = liquid and S = solid.

[b] Soil pH less than 7.

[c] Soil pH greater than 7.

[d] Dash lines indicate that surface application of these materials is not the normal cultural practice.

[e] H = loss over 40 percent; M = losses between 20 to 40 percent; L = losses between 5 to 20 percent; and V = losses less than 5 percent.

Table 10-2. Nitrogen recoveries and ammonia losses for different fertilizer materials[a].

Fertilizer material	Method of application[b]	Ammonia Nitrogen recovery	Ammonia Nitrogen loss
		------%------	------%------
Ammonium sulfate (AS)	Dry	41.9	42.3 c[c]
	Wet	31.1	53.1 d
	Liquid	23.3	60.9 e
Diammonium phosphate (DAP)	Dry	47.3	36.9 c
	Wet	27.8	56.4 d
	Liquid	24.5	59.7 d
Monoammonium phosphate (MAP)	Dry	65.1	19.1 c
	Wet	51.5	32.7 d
	Liquid	37.0	47.2 e
Urea	Dry	62.9	21.3 c
	Wet	73.2	11.0 c
	Liquid	63.9	29.3 c
Ammonium nitrate	Dry	58.4	25.8 d
	Wet	72.0	12.2 c
	Liquid	61.7	22.5 d
70% AS + 30% MAP	Dry	51.1	33.1 c
	Wet	47.6	37.6 c
	Liquid	42.6	41.6 c
70% Urea + 30% MAP	Dry	54.2	30.0 c
	Wet	62.0	22.2 c
	Liquid	51.6	32.6 c
Potassium nitrate[d]	Dry	84.2	0

Source: Adapted from Fenn and Escarzaga, 1976. Copyright © 1976 by Soil Science Society of America. Reprinted by permission.

[a] A greenhouse study using a Harkey silt loam at pH, and barley to measure nitrogen recovery.

[b] Nitrogen at 275 kg/ha was applied on the soil surface as: Dry = dry material to air dry soil; Wet = dry material to a saturated soil; Liquid = applied as an aqueous solution and left seven days before wetting the soil and planting.

[c] Means within fertilizer materials followed by the same letter are not significantly different at the 5 percent level of probability.

[d] Potassium nitrate was used as the basis for calculating recoveries and losses.

concentration at which nitrates may become toxic to plants. Very high concentration of nitrates may be found in the conductive tissue of plants. High nitrate concentration is seldom found in leafy tissue of plants. Nitrate accumulation in the leaf does occur when the nitrate reductase system is not functioning well, creating phytotoxicity in the leafy tissue. It has been the author's (RSR) experience that where molybdenum (an essential cofactor for nitrate reductase) deficiencies occur, nitrate concentrations in leaf tissue can accumulate to nearly the same levels as are found in the conductive tissues of the same plant. Plant symptoms associated with leaf nitrate accumulation have been the appearance of chlorotic spots randomly scattered about the leaf, with some tendency to occur in higher numbers at the leaf margin. As the nitrate concentration increases, these chlorotic areas coalesce and become necrotic. These symptoms generally occur on the most recently matured leaves.

Problems associated with different sources of nitrogen are the specific toxicity of the ammonium ion and the osmotic affect of fertilizer salts when these materials are placed in the proximity of seeds or roots of plants (Table 10-3). Triple superphosphate and phosophoric acid reacts rapidly with soils to form insoluble precipitates, thereby resulting in very low burn hazard. The burn hazard associated with ammonium nitrate is principally the osmotic effect. Aqua-ammonia and "10-34-0" materials have a very high burn hazard as a result of the specific ion toxicity and high soil pH when placed in the proximity of plant roots.

Another specific ion affect is the preferential absorption and utilization of different forms of nitrogen. Cox and Reisenauer (1973) demonstrated that small amounts of ammonium along with nitrate can result in a greater total nitrogen uptake by wheat. Investigations by Rauschkolb (1968) showed a preferential absorption of nitrate only after cotton became severely deficient. Many other investigators have shown that most plants are capable of utilizing ammonium or nitrate forms of nitrogen equally well. It is only in very unique conditions that preferences have been demonstrated.

Still, nitrate is frequently assumed to be the preferred form of nitrogen for plant uptake. As previous discussion has shown, nitrogen transformations in soils lead to the ultimate formation of nitrate. Thus, uptake is not so much a function of preferential absorption but rather the fact that nitrate is the principal available form of nitrogen that plants find present in the soil environment.

Some investigators have demonstrated, under some soil conditions, a difference in nitrogen-use efficiency related to the source of nitrogen used. An example of these effects is shown in Table 10-4. Ammonium sulfate and urea are readily available nitrogen sources, while sulfur-coated urea and Nitroform are so-called slow-release materials. At the same nitrogen rate the slow-release materials generally produced lower yields. Ammonium sulfate showed some advantage over urea. The advantage could be attributed to a sulfur response or the effect of the sulfate anion in mitigating an alkaline pH shift caused by the ammonium cation.

The impact of irrigation method on the selection of nitrogen source is related to the pattern of water and nitrogen movement. Patterns for movement of different nitrogen forms as a function of irrigation method and placement have been

Table 10-3. Comparison of burn hazard for commonly used fertilizers on vegetables.

N-P-K Content ---%---	Common Name (Physical State)	Ammoniacal Nitrogen ---%---	Nitrate Nitrogen ---%---	Burn Hazard[*]
83-0-0	anhydrous ammonia gas	100	0	very high
20-0-0	aqua ammonia (liquid)	100	0	very high
10-34-0	"10-34" (liquid)	100	0	very high
18-46-0	"18-46" (solid)	100	0	very high
46-0-0	urea (solid)	100	0	high
32-0-0	UN32 (liquid)	75	25	high
11-48-0	"11-48" (solid)	100	0	moderate-high
33.5-0-0	ammonium nitrate (solid)	50	50	moderate-high
20-0-0	ammonium nitrate (liquid)	50	50	moderate-low
17-0-0	"CAN 17" (solid)	25	75	low
15.5-0-0	calcium nitrate (solid)	0	100	very low
0-45-0	triple superphosphate (solid)	----No nitrogen----		very low
0-52-0	phosphoric acid (liquid)	----No nitrogen----		very low

Source: Mayberry, 1977.

[*] Relative evaluation based upon field observations and controlled research investigations by Mayberry in Imperial Co., California, on calcareous soils.

Table 10-4.   Effect of nitrogen source and rate on yield of cantaloupe, potato and tomato.

Crop	Nitrogen rate	Ammonium sulfate	Urea	Sulfur coated urea	Nitroform
	kg/ha	-------------------------yield in mt/ha-------------------------			
Cantaloupe	0	30.7[*] a[b]			
	67	44.6 c	43.6 c	37.3 b	33.6 a
	134	49.4 d	44.8 c	41.8 c	37.1 b
Potato	0	20.4[*] a			
	67	38.0 d	33.9 de	31.8 de	28.1 b
	134	44.7 h	39.1 fg	36.6 ef	30.7 bc
	301	46.6 h	43.9 h	40.9 g	36.8 f
Tomato	0	40.0[*] a			
	112	49.3 c	50.3 c	-[c]	44.4 b
	224	48.8 c	51.6 c	-	47.9 bc

Source: Adapted from Lorenz et al., 1972.

[*] These yield levels were the control yields for each experiment and are reported under a particular N source for convenience only.

[b] Means with the same letter are not statistically different at the 5 percent probability level within the values reported for each crop.

[c] Dash indicates no data.

shown in Figs. 4-3 to 4-10 and Fig. 7-5. The depth and direction of movement depends upon the uniformity of water application and the amount of water applied. As discussed previously, ammonium moves very little from point of contact with the soil. Urea is mobile in the soil until it becomes hydrolyzed and nitrate moves readily with the wetting front.

Method of application also determines the selection of nitrogen source. For example, ammonia volatilization losses have been measured when ammonium is applied by flood or sprinkler irrigation application techniques. However, when applied in drip irrigation, the loss is reduced, since there is mass flow of ammonium into the soil profile and it is applied to a much smaller area of soil. Because urea and nitrates are more mobile in soils and not as susceptible to ammonia volatilization losses, they are more widely adapted for use with different irrigation methods and soil conditions.

Some ammonium sources when applied in irrigation water will increase the pH of the water to the point where calcium carbonates will precipitate. This has the potential for causing plugging of sprinkler and drip irrigation lines when the water contains dissolved calcium and carbonate. If the pH is maintained at neutral or lower, the possibility of precipitation and volatilization losses are reduced. Figure 10-1 shows the magnitude of ammonia volatilization loss when it is applied through a sprinkler system. The loss is shown as a function of a change in concentration or pH of the irrigation water.

The impact of different irrigation methods on the movement of nitrogen in

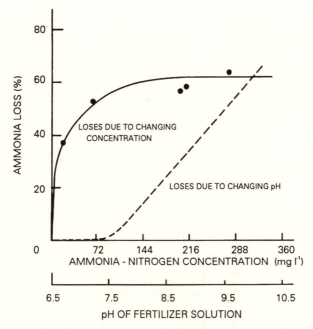

Figure 10-1 Ammonia loss from irrigation sprinkler jets. (Henderson et al., 1955. Copyright © 1955 by the American Society of Agricultural Engineering. Reprinted by permission.)

the soil profile where different sources of nitrogen and methods of placement have been utilized are shown in Figs. 10-2 to 10-5 (see also Fig. 7-5). Soil solution extracts were obtained after a 3-year period of irrigation and cropping with the treatments indicated in the figure legends. In the furrow-irrigated plots the nitrate-N concentration appeared to be slightly greater than URAN (urea + NH_4NO_3) rather than ammonium was the source of nitrogen. Under subirrigation and sprinkler irrigation, there seems to be little treatment difference due to source except where anhydrous ammonia was banded in the bed and sprinkler irrigated. Earlier discussions indicated that banding of ammonia inhibits the rate of nitrification, which reduces the leaching potential. Coupled with a lower leaching fraction under sprinkler irrigation this resulted in accumulation of nitrate in the upper portions of the soil profile. The small difference in the subirrigated plots between nitrogen sources is attributed to the effect of denitrification on nitrate levels.

The nitrogen recoveries and ammonia volatilization losses for different nitrogen sources when applied to soil at different water contents are shown in Table 10-2. All nitrogen sources had significantly lower recoveries than the potassium nitrate standard. Differences were also observed in method of application for some of the sources evaluated. These data serve to show the importance of the method of application and selection of a particular source of nitrogen for certain conditions in order to enhance efficiency of utilization.

Manure is an important nitrogen source. Differences in the nitrate nitrogen

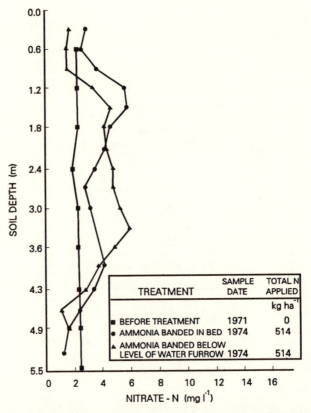

Figure 10-2 Soil extract (1 : 1) nitrate-N concentrations by soil depth before treatment (1971) and after four years of nitrogen application and furrow irrigation (1974) in a Miles fine sandy loam soil (fine-loamy, mixed, thermic udic Paleustalfs.) (Wendt et al., 1977.)

concentration is soil-water as a function of different rates of manure application in either liquid or solid form are shown in Table 10-5. Higher nitrogen concentrations were found in the soil profile with liquid manure sources for comparable rates of liquid and solid manure. The liquid manure has a much higher concentration of readily available nitrogen, which contributes to the greater concentration of nitrate-N in the soil profile. With a slightly higher water application rate the soil nitrate-N concentration was somewhat lower. The Ramona and Domino soils would have a greater potential for denitrification. In fact, the data in the study show that in these soils denitrification losses were substantially greater than for the Hanford, thus reducing the nitrate leaching potential. However, the loss due to both denitrification and volatilization losses was substantially greater than for the Hanford, thus reducing the nitrate leaching potential. In the Hanford soil, the loss due to both denitrification and leaching was of the same order of magnitude. Another important point to make here concerns the amount of manure that is required to achieve maximum crop yields. Significantly greater yields were obtained from plots receiving manure as compared to

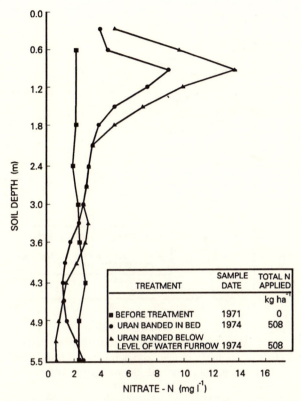

Figure 10-3   Soil extract (1 : 1) nitrate-N concentrations by soil depth before treatment (1971) and after four years of nitrogen application and furrow irrigation (1974) in a Miles fine sandy loam soil (fine-loamy, mixed, thermic udic Paleustalfs.) (Wendt et al., 1977.)

the control plot; however, there were no differences in yield between plots receiving different rates of manure. As with inorganic sources of nitrogen, any amount of manure applied in excess of that required to achieve maximum yield leads to increased nitrate pollution potential.

The effect of the environmental factors on the selection of nitrogen source is related primarily to the effect of temperature on nitrogen transformations in the soil and the influence of precipitation on nitrogen movement. The effect of temperatures is principally on the rate of biological processes. Data by Frota and Tucker (1972) showed a slight effect of temperature on uptake of N from different sources. More ammonium than nitrate was taken up by lettuce at cooler temperatures, but for either source the rate was not adequate to meet the growth rate of the plant. It was unclear from these data whether the growth rate of the plant, which is slowed by cool temperatures, is the controlling factor of nitrogen uptake or whether there is some transport mechanism at the root surface that limits the rate of absorption. In either case, the net effect is a reduced rate of nitrogen uptake at cool temperatures. The effect of precipitation is to reduce the potential for ammonia volatilization losses by moving the ammonium fertilizers

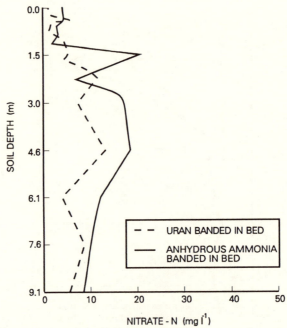

Figure 10-4 Porous bulb soil-water extracts showing nitrate-N concentration with depth after 3 years of applied nitrogen and subsurface trickle irrigation in a Miles fine sandy loam soil (fine-loamy, mixed, thermic udic Paleustalfs.) (Wendt et al., 1977.)

into contact with the soil where it becomes absorbed on soil particles. While volatilization losses may be reduced in calcareous soils under these conditions, the losses are still relatively large. In the case of nitrate and urea the amount of leaching could be increased in coarse textured soils, because these compounds can move through the soil profile with the wetting front.

The fate of different sources of nitrogen using various combinations of placement and irrigation are shown in Fig. 10-6. Losses and movement of the different nitrogen sources are shown under the treatments indicated. Urea applied to the surface of a moist soil will absorb moisture from the air, allowing the urea to come into contact with the soil and undergo hydrolysis. The ultimate formation of ammonium hydroxide and calcium sulfate, which precipitates, causes a dramatic pH shift in the vicinity of the fertilizer particle, creating a condition conducive to ammonia volatilization. The ammonium retained in solution has a positive charge that allows it to be adsorbed by soil particles. The equilibrium between the absorbed and solution ammonium then controls the amount of ammonia volatilization loss. The ammonium in solution is then subject to transformation to nitrate. Nitrate may then be moved into the soil profile with irrigation or precipitation. Ammonium sulfate acts similarly when broadcast on the soil surface. Ammonium nitrate applied to the surface also absorbs moisture from the air, causing its dissolution and allowing it to come in contact with the soil. However, because of the high solubility of calcium nitrate, a by-product of

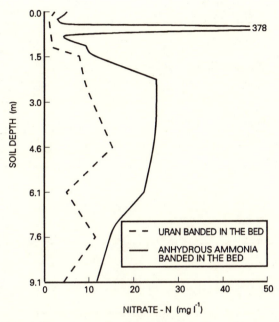

Figure 10-5 Porous bulb soil-water extracts showing nitrate-N concentration with depth after 3 years of applie dnitrogen and sprinkler irrigation in a Miles fine sandy loam soil (fine-loamy, mixed, thermic udic Paleustalfs.) (Wendt et al., 1977.)

Table 10-5. Mean NO₃-N concentrations for the water in the 1.5- to 4.5-m depth after 4 years of manure application.

| Soil | Manure rate[a] | Average NO₃-N concentration in soil water | |
		Low irrigation (114 cm)	High irrigation (142 cm)
	mt/ha/yr	---------------------mg/l---------------------	
Hanford	0	16	21
	21 (L)[b]	85	70
	40 (S)	51	52
	42 (L)	170	116
	79 (S)	131	83
	158 (S)	190	161
Ramona	79 (S)	59	47
Domino	79 (S)	53	24

Source: Pratt, 1979a.

[a] Manure application rates expressed on air-dry weight basis.

[b] L = feedlot manure collected under slatted floors and applied as a slurry.
    S = feedlot manure collected under slatted floors, air dried and applied as a solid.

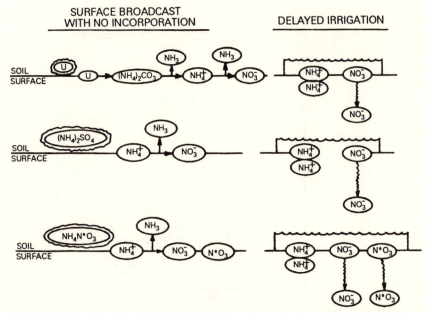

Figure 10-6 Schematic showing volatilization loss, transformation, and movement for nitrogen from various sources when broadcast on the surface and subsequently irrigated or when applied in irrigation water. The asterisk indicates the fate of the nitrate ion present in the fertilizer source as nitrate-N.

the hydrolysis, the soil pH shift to a more alkaline condition does not occur. Consequently, ammonia volatilization losses are minimized. The remaining ammonium ion acts as described before.

When nitrogen fertilizer materials are added in the water, the distributions and susceptibilities to loss are different. Ammonium is subject to losses as indicated in Fig. 10-6. The ammonium that remains in solution can move slightly into the soil profile (Fig. 10-5). Once the ammonium comes into contact with soil, it undergoes the reactions described above. Because of its initial zero net charge, urea moves readily into the soil profile. Urea then undergoes the same sequence of events as described for the surface application of urea. However, because these processes occur beneath the soil surface, the potential for volatilization loss is essentially eliminated. When ammonium nitrate is applied in irrigation water, the ammonium moves slightly into the soil profile and the nitrate moves along behind the wetting front. During a drying period, nitrification of ammonium occurs; a subsequent irrigation moves the nitrate farther into the soil profile.

The fertilizer source plays a major role in the type of placement that may be achieved, the magnitude of the losses that might occur, the nitrogen uptake efficiency, and the nitrate pollution potential. It should be clear that nitrogen fertilizers cannot be treated alike. To insure the greatest efficiency one has to

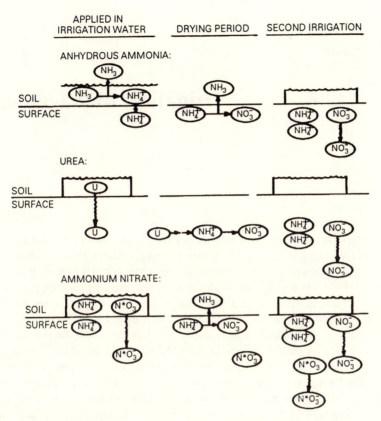

Figure 10-6 (Continued)

select the source that is best adapted to the conditions under which it will be applied.

11

Irrigation Management

Management of irrigation water influences the movement of nitrogen in soils, and thus the ability of the plant to utilize nitrogen, which ultimately effects the efficiency of nitrogen use. In order to manage water for efficient fertilizer use, an understanding is required of the differences between irrigation methods with respect to rates of application, timing, the uniformity of application, and frequency of application which interact with soil nitrogen to influence leaching, denitrification, and positional availability of nitrogen. These factors are also influenced by soil physical characteristics. In Fig. 11-1 are shown the depths of water penetration and the amounts of available water per unit depth in the soil profile for different soil textures. Coarse textured soils have a greater potential for leaching losses because of the greater depth of water movement for a given amount of surface-applied water. For the same amount of surface water applied, the depth of water penetration is considerably less in a fine textured soil. The effect of soil structure and texture on pore size distribution and ultimately on the infiltration rate and internal drainage causes differences in the movement and potential losses of nitrogen that occur as a function of the application of water.

Soil uniformity and depth may also determine nitrogen availability and utilization. In highly nonuniform soils there are great differences in the rate of infiltration, leading to wide variation in the potential for leaching or denitrification. If a field is irrigated to provide adequate water for a fine textured soil, then excessive leaching occurs in the portion of the field where the soils are coarse textured. Conversely, if the field is irrigated to supply adequate water in the coarse textured soil, the amount of water applied to the fine textured soil results in nearly saturated conditions, creating a high potential for denitrification losses. The depth of the soil determines the frequency of irrigation and the capacity of the soil to supply adequate water for the plant. A shallow soil would have to be irrigated frequently in order to provide adequate water for a plant. Frequent irrigation on a shallow soil could lead to overirrigation and increase the potential for development of a perched water table, and thus the potential for denitrification. Knowledge of the uniformity and depth of the soil along with the type of plant root system is essential for proper water management in order to avoid excessive applications that may lead to high leaching or denitrification losses.

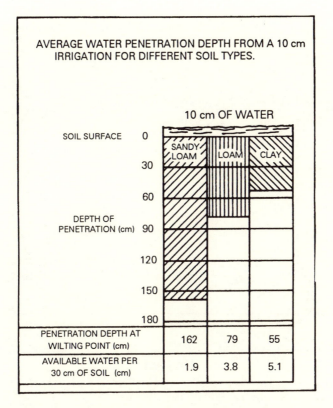

Figure 11-1 Schematic showing depth of water penetration after irrigation of different soil types at their permanent wilting point. (Luthin, 1970.)

Crop characteristics that influence irrigation management are evapotranspiration and the depth or extent of their rooting system. Plant extraction of water roughly corresponds to 40, 30, 20, and 10 percent of the water available from each successively deeper one-quarter of the root zone. However, this does not recognize the development of plant roots with time. Figure 11-2 shows the development of a corn root over the growing season, showing that the center of water extraction by roots occurs deeper in the soil profile as the plant matures for the special case where additional water is not applied to the field.

The frequency with which water is applied and the amount of water required to wet the soil profile in order to provide adequate water for the plant to the bottom of the root zone are determined to a large extent by the rate of development and the ultimate extent of the root zone. Shallow rooted crops must be irrigated more frequently, thus creating the potential for more nitrogen loss. Compacted layers in the root zone can restrict root development, thus influencing extraction patterns and management practices.

Seasonal evapotranspiration, as shown in Table 11-1 for several different crops, influences irrigation management. Wide ranges in the amount of evapotranspiration result from length of growing season, the percentage of ground cover, and the genetic characteristics of a plant. Also shown in Table 11-1 are

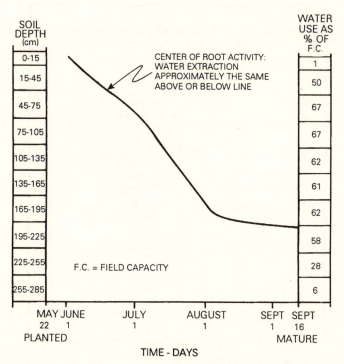

Figure 11-2   Root extraction of water as a function of time and depth from a non-irrigated plot of corn (*Zea mays*). (Stewart, 1971.)

estimates of irrigation application efficiencies that were commonly found in the area. With an estimate of the evapotranspiration for the season and the irrigation application efficiency, one can determine the seasonal need for water and use this as a guide for irrigation management. Potential nitrogen losses are greatest where the evapotranspiration is high and the application efficiency is low, thus creating the condition where large amounts of water would be applied in order to maintain the plant. With relatively high application efficiencies the potential for leaching or denitrification losses is reduced, regardless of the evapotranspiration potential.

The irrigation method, insofar as it determines the uniformity, amount, and application efficiency, plays an important role in determining the irrigation management for obtaining the greatest nitrogen-use efficiency. The coefficient of uniformity determines to a large extent the efficiency with which water is applied to a given field. The impact of changes in the coefficient of uniformity on the amount of water required to irrigate a given area is shown in Table 11-2. The higher the coefficient of uniformity, the less water is required to achieve a given application rate. The water saved by improving the coefficient of uniformity is available for an alternative use. One of the obvious benefits of this improvement in coefficient of uniformity is the impact on application efficiency and subsequently the reduction of the leaching losses that might occur from the root zone in a particular management area. For example, Fig. 11-3 shows that

Table 11-1. Estimated evapotranspiration (ET), typical application efficiency, and calculated seasonal applied water requirements for various crops.

Crop	ET	Typical application efficiency	Seasonal needs of applied water
	----cm----	------%------	------cm------
Alfalfa	101.6	65	156.2
Almonds	88.1	85	103.6
		65	135.1
Beans	44.5	65	68.3
Cereal grains	30.5	60	50.8
Cherries	92.7	85	109.0
Corn	64.0	60	106.7
Grapes	53.1	60	88.4
Pasture	111.8	50	223.5
Peaches	91.4	65	140.7
Sudan	45.7	60	76.2
Sugar beets	71.7	60	118.6
Tomatoes	64.8	60	108.0
Walnuts	109.7	85	129.0
Watermelons	45.7	55	83.1

Source: Adapted from Anonymous, 1977.

Table 11-2. Effect of coefficient of uniformity on water requirements[a]

Uniformity coefficient (CU)	Average application required over the area to apply 5 cm on 75% of area	Water saved by improving CU from 70%	Increase in area that can be irrigated by improving CU from 70%
-----%-----	------cm------	------%-----	------%------
90	5.56	18.3	22.4
80	6.12	10.1	11.2
70	6.81	0.0	0.0

Source: Adapted from Shearer, 1969.

[a] Calculated from normal distribution curves.

nitrate-N leaching is greater where the length of time of the irrigation set is the greatest. While the nitrate concentrations at the 105-cm soil depth were not very different between the two regimes, the total magnitude of loss was much greater for the treatment applying excessive water.

Nitrate distribution associated with different leaching fractions is shown in Fig. 11-4. With a relatively small leaching fraction, the nitrate-N remains at a

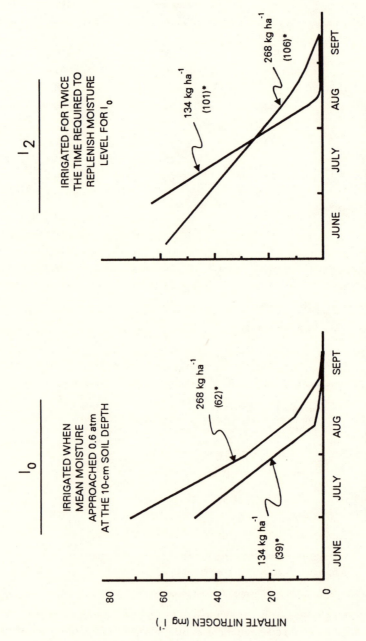

Figure 11-3 Concentration of nitrate-N at the 105-cm soil depth for two irrigation regimes (I_0 and I_2) for two rates applied nitrogen. Values in parentheses are estimated kg/ha of nitrogen leached below the 105-cm soil depth. (Jensen et al., 1965.)

relatively high concentration in active root zone. With progressively higher leaching fractions, the nitrate-N concentration in soil profile declines. These data illustrate the importance of leaching fraction in determining the pollution potential of irrigation management practices. Contrasting nitrate concentration distributions in Fig. 11-4(A) and 11-4(B) with that in Fig. 11-4(C) shows that mass emission of nitrogen was greater with the higher leaching fraction. This is evident from the low residual nitrogen shown in the root zone (Fig. 11-4(C)).

Denitrification potential is enhanced when the soils are kept very moist (Rolston et al., 1977). Based on this information, it can be concluded that frequent light irrigations would tend to create a condition where greater denitrification would occur. Other evidence developed by Smika et al. (1977) shows that infrequent heavy irrigation increases leaching losses substantially, especially on sandy soils. Information from these and other studies on frequency and amount of irrigation under specific conditions provides clearer guidelines as to the method of nitrogen management that will provide greater nitrogen-use efficiency for a given soil and irrigation method.

The source and rate of nitrogen will also influence the magnitude of leaching losses and in some respects determine the type of irrigation management for obtaining the greatest efficiency. For example, in Table 11-3 is shown the influence of soil-water content and the amount of applied water on the depth of movement of surface-applied ammonium added as ammonium sulfate. The depth of ammonium movement into the soil profile increased with depth of water applied. The depth of ammonium movement is also greater for any given amount of applied water in initially wet soils than when the soils are initially dry. This shows the impact of the interaction between irrigation management and fertilization on nitrogen-use efficiency for surface-applied fertilizers. Even though most of the ammonium penetrated 5 to 7 cm into soil profiles, substantial ammonia volatilization losses still could be measured (Fenn and Escarzaga, 1976).

The interaction between the rate of fertilizer and amount of water applied is also a factor in determining the efficiency of nitrogen utilization. In Fig. 11-5 is shown the net removal and net addition of nitrogen to corn. Where maximum grain yield occurred there was a net removal of nitrogen by corn grain and stover in 1976. Similar information has been developed by Broadbent and Carlton (1976) for other years in the same plots. As the amount of applied nitrogen exceeded that required for maximum yield, there was a net nitrogen addition to the soil, which had the potential for leaching. The impact of applied nitrogen rate and amount of irrigation on the uptake of soil nitrogen and carry-over are shown in Table 11-4. There was a continuous decline of soil nitrogen uptake in the control plots over time. This was not observed at the rate of applied nitrogen that produced maximum yield (90 kg/ha of N in 1974 and 180 kg/ha in 1975). There was relatively little difference in the amounts of residual nitrogen where low or adequate amounts of water and nitrogen were applied. Where relatively large amounts of nitrogen were applied and at the highest irrigation level, there was a substantial decrease in the residual nitrogen in the soil profile. This is additional evidence of the high mass emission of nitrogen for combinations of high rates of fertilization and excessive irrigation.

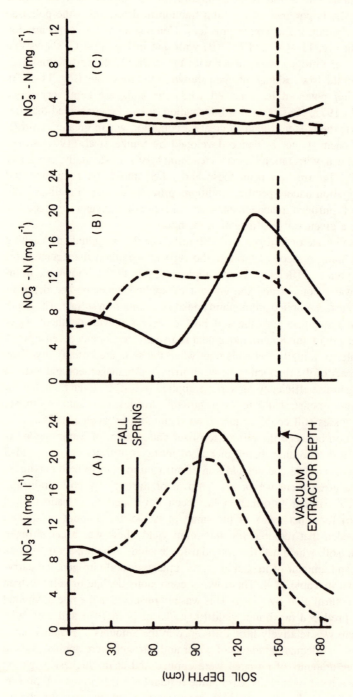

Figure 11-4 Three-year average of nitrate-N distribution for spring and fall in a loamy fine sand soil with (A) 2.5%, (B) 4.8%, and (C) 10.2% of applied water percolating to the 150-cm soil depth. (Smika et al., 1977. Copyright © 1977 by the American Society of Agronomy. Reprinted by permission.)

Table 11-3. Movement of NH_4^+-N into silty clay loam soil with increasing increments of water.

Initial water content	Water added	Soil depth increments, cm					
		0-2.5	2.5-5	5-7.5	7.5-10	10-12.5	12.5-15
	cm	------------------------% of NH_4-N Added*------------------------					
Saturated	2.5	38	28	10	5	5	5
Dry	2.5	60	27	6	1	< 1	---
Saturated	5.0	42	22	29	13	13	< 1
Dry	5.0	48	30	14	1	1	< 1
Saturated	11.2	35	33	23	6	3	2
Dry	11.2	40	26	17	8	4	2
Saturated	22.5	29	21	17	11	10	7
Dry	22.5	22	27	20	14	8	5

Source: Fenn and Escarzaga, 1977. Copyright © 1977 bt Soil Science Society of America. Reprinted by permission.

* NH_4^+-N added at 550 kg N/ha.

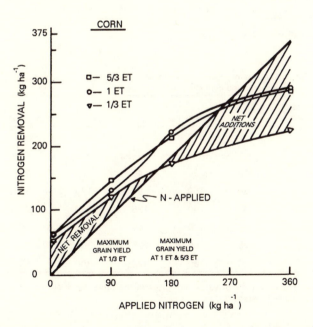

Figure 11-5 Influence of amount of applied nitrogen and water on the net crop removal by corn (*Zea mays*) grain and stover and carry over of nitrogen in the Yolo Loam soil (fine-silty, mixed, nonacid, thermic typic Xerorthents) for the Davis site, 1976. (Broadbent and Carlton, 1976.)

Table 11-4.   Uptake of soil nitrogen by corn and residual inorganic nitrogen in the three meter soil profile as affected by irrigation amount and fertilizer rate, Davis site.

	Nitrogen applied, kg/ha/yr							
	0		90		180		360	
Year	Uptake	Res.	Uptake	Res.	Uptake	Res.	Uptake	Res.
	--kg of N/ha---							
20 cm Irrigation water								
1973	155	139	139	225	110	267	95	379
1974	83	119	99	147	93	194	97	356
1975	64	110	102	133	105	184	107	558
60 cm Irrigation water								
1973	155	110	148	238	108	261	84	377
1974	120	120	134	117	115	162	87	323
1975	90	120	108	138	124	139	82	409
100 cm Irrigation water								
1973	145	126	136	146	103	176	75	381
1974	105	137	133	116	117	154	82	333
1975	77	117	100	138	101	134	83	295

Source: Broadbent, 1976.

Table 11-5.   The amount of nitrogen applied to soil as a function of quantity of irrigation water and concentration of nitrogen.

Hectare - meters of irrigation water	Parts per million of nitrogen in irrigation water					
	5	10	15	20	25	30
	---kg of N/ha[a]---					
0.3	15	30	45	60	75	90
0.6	30	60	90	120	150	180
0.9	45	90	135	180	225	270
1.2	60	120	180	240	300	360
1.5	75	150	225	300	375	450
1.8	90	180	270	360	450	540

[a] To calculate values in the table or values not given in the table; kg of N/ha equals (ppm N x ha-m of applied water).

A frequently overlooked aspect of irrigation and nitrogen management is the contribution of nitrogen in the water supply to the amount of available nitrogen for plant growth. In Table 11-5 are shown the amounts of nitrogen applied per hectare calculated for different volumes of irrigation water at various nitrogen concentrations. If one hectare-meter of water were required to meet the consumptive use of the plant, this would contribute approximately 100 kg/ha of N if the nitrate-N concentration were 10 ppm. This amount of added nitrogen is not adequate to produce maximum yield for most crops. Surface water supplies are usually very low in nitrate-nitrogen, ranging between less than 1 ppm to as much as 5 ppm nitrate-N, if there is no contamination from domestic effluents, subsurface return flows, or surface return flow in which nitrogen fertilizer has been picked up. Groundwater supplies, on the other hand, may range from very low to very high concentrations of nitrate-N. At high nitrogen concentrations it is possible to see an adverse effect on yield and quality of crops that are sensitive or even moderately tolerant to excessive nitrogen.

Irrigation management must also take into account the environmental effects such as temperature and relative humidity as well as precipitation frequency and amount. Rainfall distribution in irrigated agricultural areas may range from relatively uniform distribution over the entire year to very seasonal distribution. Most of the irrigated areas of the United States have such high potential evapotranspiration that sparse rainfall even during the growing season is frequently not considered in the water budget. In other irrigated areas the distribution of rainfall during the season is not satisfactory to provide adequate water during critical periods of crop growth and development. Consequently, irrigation is practiced in order to prevent water stress during these critical periods and thereby maximize production. There are several irrigation scheduling techniques available that utilize combinations of solar radiation, relative humidity, temperature, and soil water-holding capacity to determine the consumptive use of water by plants and predict the timing and amount of irrigation to replenish the soil-water supply. Since these factors play a role in governing the amount of water applied, they influence the magnitude of nitrate leaching, ammonia volatilization, and denitrification losses that might occur. By precisely predicting the timing and amount of irrigation, it is possible to mitigate somewhat the magnitude of such losses by virtue of minimizing the application of excess water.

In an irrigated agricultural system it is not practical or desirable to prevent leaching of water below the root zone of the crop. This is essential for maintenance of salt balance in the soil. As a result some leaching of nitrogen does occur. Through irrigation and nitrogen management, the objective is to use both in a manner with crop production and protection of the environment. A corollary goal is to reduce mass emission of nitrogen below the root zone. Concentration of nitrogen in the soil solution alone is not a good measure of pollution potential.

12

Timing of Nitrogen Application

The timing of nitrogen fertilizer applications is one of the key management techniques available to a grower for improving the efficiency of nitrogen use. Proper timing of applications to enhance the efficiency of nitrogen utilization requires an understanding of how plant characteristics determine the need for nitrogen during the season. Environmental factors regulate the availability of nitrogen by their influence on transformation rates, plant uptake, and conditions conducive to nitrogen losses. There is also considerable interaction between timing and placement. Appropriate timing is defined as having nitrogen positionally available in the quantity needed in time to meet the demand of the plant for the nutrient. In order to ascertain the appropriate timing of fertilizer nitrogen applications it is necessary to understand the effect of the system variables on the characteristics outlined above. The closer one can adjust the application of nitrogen to insure its availability to the plant when required, the greater the efficiency by virtue of enhanced opportunity for plant absorption before it is exposed to the myriad of loss mechanisms that decrease its availability to the plant.

Soil characteristics that have an influence on the timing of nitrogen fertilizer applications are primarily structure and texture of the soil. Both influence the infiltration rate of water and the aeration of the soil, which in turn affects the denitrification and leaching potentials of nitrogen. These soil characteristics also influence the susceptibility of soils to compaction and thereby development of the root systems of plants, which have an indirect effect on the positional availability of nitrogen to the plant. If the root system is unable to develop adequately, there is not sufficient root volume to supply the amount of nitrogen required for plant growth and development. Furthermore, the root system may be prevented from reaching the nitrogen supply because of improper mechanical placement of the fertilizer or if nitrogen movement by water is away from the root zone.

The effects of soil texture and timing of fertilizer nitrogen on yields are shown in Table 12-1. Even with the more efficient side-dress method of application, the yield of corn on the clay soil was still less than the yield obtained on the loam with a fall application of nitrogen. These data suggest that soil texture has an effect on such factors as root aeration and denitrification potential. The lower

Table 12-1.   Effect of time of nitrogen application on yield of corn grain on clay and loam soil.

Soil type	Time of N application		
	Fall	Spring	Side-dress
	----------------------------Yield, kg/ha*----------------------------		
Clay	5,680	6,970	7,160
Loam	7,730	8,400	8,490

Source: Adapted from Stevenson and Baldwin, 1969. Copyright © 1969 by American Society of Agronomy. Reprinted by permission.

* Based on 15.5% moisture: Average yield for the unfertilized treatment was 5,140 kg/ha.

yields from the fall-fertilized plots as opposed to those side-dressed during the growing season are attributed to the denitrification and volatilization losses that occurred between the time of application and the time when nitrogen was being absorbed by the plant. The higher yields on the same soil for the same rate of nitrogen at all times of application are attributed to a much lower denitrification potential for the loam soil. As indicated in Table 12-2, soil type also has an impact on the rate of ammonium transformation to nitrate at the lower temperature. These types of variations are found to exist but are not readily explained without extensive soil characterization. When one is aware of these differences, they should be taken into account in determining the appropriate timing of nitrogen fertilizer applications.

Perhaps one of the most important determinants of the timing of fertilizer applications is the plant requirement for nitrogen during the season. In Chapter 3, the relationship between dry matter accumulation and plant nutrient uptake was shown (Figs. 3-17 to 3-24). Sufficient evidence exists to support the conclusion that dry matter accumulation can be used as a guide for determining nitrogen demand by the plant. There is still the need to develop more detailed

Table 12-2.   Comparison of maximum nitrification rates of aqua ammonia at two temperatures for different soils.

Soil	Nitrogen Added	Period of maximum rate in weeks		Maximum rate in kg N/ha/day	
		7°C	24°C	7°C	24°C
	----ppm----				
Sacramento	50	0-2	0-1	8	25
clay loam	100	0-2	0-1	9	25
	200	0-2	0-1	9	27
Yolo loam	50	2-4	0-1	5	20
	100	4-6	0-1	9	29
	200	4-6	0-1	15	38

Source: Adapted from Tyler et al., 1959.

information for more crops in order to more precisely determine the nitrogen demand and thereby establish guidelines for fertilization. Plant nitrogen demands are different among plant species (Figs. 3-17 to 3-24). Corn and lettuce are two very different examples showing the importance of knowing the seasonal demand for determining when to apply nitrogen fertilizer. The impact of such timing on nitrogen-use efficiency can be seen in Table 12-3. Where nitrogen was applied at the rate of 112 kg/ha of N, the yield of lettuce was just as great or greater than when 224 kg/ha of N was applied. The nitrogen carryover (post-harvest) was reduced at the lower rate and when fertilizer was not applied after the folding stage of plant development. Approximately 80 percent of the plant nitrogen was absorbed between folding stage and harvest. In this case late and/or excessive rates of nitrogen did not have an adverse effect on yield or quality. However, plants that are sensitive to excessive nitrogen fertilizer can have their yield or quality adversely affected.

Another plant characteristic that determines the suitability of split applications and different timing of fertilizer applications is the rate of decrease of nitrate-N concentration in the tissue with time and the ability of nitrogen applied at some period during plant growth and development to be absorbed in sufficient quantities to avoid a deficient period that has an adverse effect on yield. Figure 12-1 shows the effect of different timing of fertilizer nitrogen applications on cotton yield. The split application resulted in a greater yield than a single application at planting, implying an increased efficiency. Grains exhibit similar changing nitrogen concentrations in their conductive tissue, but at a characteristically lower nitrate-N level.

The relationship between yield and timing of application of ammonium nitrate for Anza wheat is shown in Fig. 12-2. These data demonstrate that increased yields and nitrogen-use efficiency occur when fertilizer applications match the temporal nutrient demands of the crop. As in this case, yield will be reduced when plants become deficient. However, when a deficiency occurs a greater yield can be obtained when nitrogen is applied than when it has not been applied. The results observed in Fig. 12-2 also demonstrate the advantage of the proper timing of nitrogen fertilizer application to achieve maximum yield. The magnitude of the yield difference for slightly different timing of nitrogen application indicates how critical timing sometimes can be. Widely varying responses of nitrogen-deficient plants to nitrogen fertilization point out the necessity of understanding the plant's growth characteristics, its critical levels, and needs for nitrogen in order to time the fertilizer nitrogen applications in such a way as to avoid yield reductions. By applying that knowledge, maximum yields can be attained with the minimum amount of nitrogen fertilizer, thus achieving high production and environmental protection.

The method of irrigation can play a role in determining alternatives for timing of fertilizer applications. Because of the different nitrogen movement with water associated with irrigation methods, different placements may be achieved; these in turn have an impact on the timing of the nitrogen fertilizer application. As indicated in previous discussion, it is possible to apply nitrogen but because of its placement render it unavailable for plant use. The method and timing of

Table 12-3. Effect of rate and timing of nitrogen application on soil nitrate-N in side of beds for three years of spring grown lettuce.

Treatment	Time of application[a]					Time of soil sampling			Heads per plot[b]
total N	PP	TH	12-L	FD	HD	PP	TH	Post harvest	
	----------kg/ha----------					----------ppm, NO$_3$-N----------			
1967									
224	56	56	--	56	56	42 b[c]	19 b	15 c	235 b
224	56	--	56	56	56	40 b	18 b	12 bc	240 b
112	56	--	--	56	--	34 b	20 b	13 c	236 b
112	56	--	--	--	56	42 b	17 b	13 c	236 b
112	--	56	--	56	--	27 a	10 a	9 ab	236 b
0	--	--	--	--	--	25 a	12 a	7 a	159 a
1968									
224	56	56	--	56	56	38 b	23 b	17 bc	254 c
224	56	--	56	56	56	41 b	26 b	20 c	249 bc
112	56	--	--	56	--	42 b	25 b	12 ab	239 b
112	56	--	--	--	56	43 b	25 b	11 ab	246 bc
112	--	56	--	56	--	29 a	7 a	15 bc	255 c
0	--	--	--	--	--	30 a	8 a	8 a	206 a

Continued--.

Table 12-3 Continued

Treatment	Time of application[a]					Time of soil sampling			
Total N	PP	TH	12-L	FD	HD	PP	TH	Post harvest	Heads per plot[b]
	-------kg/ha-------					-------ppm, NO₃-N-------			
1969									
224	56	56	--	56	56	35 b	19 c	24 cd	173 bc
224	56	--	56	56	56	32 b	19 c	25 d	170 bc
112	56	--	--	56	--	30 b	19 c	18 bc	179 bc
112	56	--	--	--	56	34 b	17 bc	16 b	153 b
112	--	56	--	56	--	22 b	12 c	17 b	188 c
0	--	--	--	--	--	23 a	14 ab	10 a	76 a

Source: Adapted from Gardner and Pew, 1974.

[a] PP, TH, 12-L, FD, and HD refer to preplant, thinning, 12-leaf, folding and heading stages of lettuce development, respectively.

[b] Number of heads per 80' x 3.3' plot (approximately 0.0049/ha).

[c] Numbers followed by the same letter in a column within each year are not statistically different at 5%.

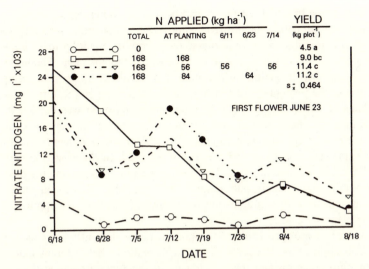

Figure 12-1 Yield of seed cotton (*Gossypium*) and nitrate-N concentration in cotton petioles as influenced by rate and time of nitrogen application. (Gardner and Tucker, 1967. Copyright © 1967 by the Soil Science Society of America. Reprinted by permission.)

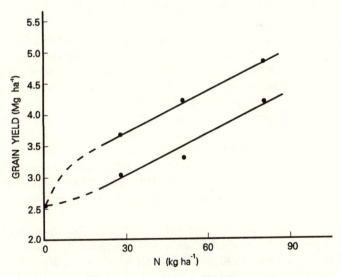

Figure 12-2 Response of Anza wheat (*Triticum aestivum*) to different timing of nitrogen application with ammonium nitrate. (Salisbery, 1981.)

irrigation can influence the timing of the nitrogen application in order to have it positionally available at the time the plant requires it.

For example, in Chapter 10 it was pointed out that nitrogen applied on the surface of the soil would not be available for plant use until such time as the irrigation water or rainfall moved the nitrogen into the root zone. If the nitrogen source is the type that will not move readily with the water, then it may require a second irrigation before the nitrogen is moved into a position to be utilized by the plant. It is possible to move nitrogen away from the plant by irrigation even though nitrogen was initially in position to be utilized by the plant. The influence of furrow and drip irrigation on the amount of fertilizer nitrogen taken up by the plant and the percentage of fertilizer nitrogen in the plant at different sampling dates is shown in Fig. 12-3. In treatment 1, nitrogen was banded in the bed and then the plants were furrow irrigated. As a result, nitrogen was moved toward the plant, allowing greater uptake of fertilizer nitrogen than for treatment 2, where nitrogen also was banded but drip irrigated. In this case, the emitters were placed between the plants and the fertilizer band, thus moving the nitrogen away from the plant. The irrigation method and selection of nitrogen source are important considerations in adjusting the timing of the fertilizer application to meet the plant demand.

The source of nitrogen is also an important factor that influences the timing of fertilizer application. Because of the transformations that different sources of nitrogen must undergo and the differences in the types of losses that might occur, the source of nitrogen is understandably a critical aspect of timing. When ammonium sources are applied to the surface, they are subject to volatilization losses as discussed previously. In order for ammonium to be available in the root zone from a surface application it must first be transformed to nitrate and moved into the root zone by either precipitation or irrigation before it becomes available for plants. Nitrate that is surface-applied is not susceptible to volatilization losses *per se*, but once it has been moved into the root zone, it can be lost through denitrification. Ammonium, on the other hand, when placed in the soil profile and prevented from undergoing transformation to nitrate, is not subject to denitrification losses except under the most extreme conditions such as are not normally found in an agricultural production system.

Climate and soils have a major impact on the suitability of different nitrogen sources for timing of application and, as indicated in Table 12-4, the method of placement can also influence the suitability of a particular timing with different sources of nitrogen. For example, injected anhydrous ammonia showed no yield difference regardless of its time of application. Spring applications of ammonium nitrate and urea showed a definite yield advantage over fall applications.

Another technique that is available to prevent the transformation of ammonium and thereby provide greater latitude for its timing is through the use of nitrification inhibitor. Yield data for winter wheat with different sources of nitrogen applied with and without a nitrification inhibitor are shown in Table 12-5. In this case, ammonium sulfate applied in the fall produced a significantly lower yield than ammonium sulfate applied in the spring or ammonium sulfate applied in the fall with a nitrification inhibitor.

Climate influences nitrogen transformations and losses, thereby affecting timing. As shown in Tables 12-1, 12-4, 12-5, 12-6, and 12-7, nitrogen recovery (as indicated by yield) is generally less for fall applications of any of the nitrogen sources. Similar information is shown in Fig. 12-4 for a fall and a spring application of urea and sodium nitrate. In every case, spring application was more

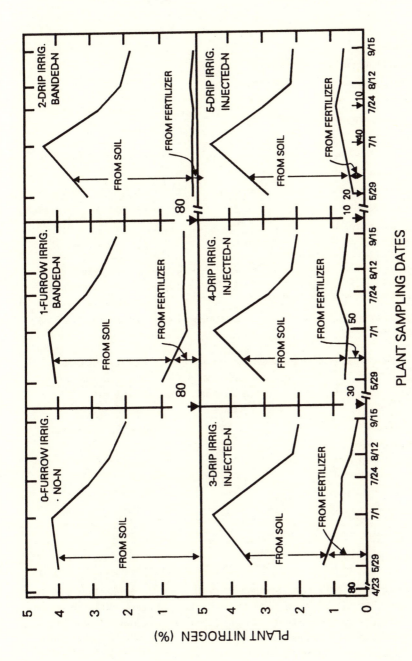

Figure 12-3 Relation between soil and fertilizer nitrogen content in tomato (*Lycopersicon esculentum*) plants at different plant sampling dates for all treatments. Numbers along abscissa indicate time and amount of fertilizer nitrogen in applied kilograms per hectare. (Miller et al., 1976.)

Table 12-4.   Effect of time of application of nitrogen fertilizer on the five year average yield of corn (Zea mays.)

Material used and method of application	Time of application	Relative average yield[a]
		--------------------%----------------
No applied N	----------	100
Ammonium nitrate:		
Surface broadcast	November	156
Surface broadcast	February	202
Surface broadcast	April	198
Urea:		
Surface broadcast	November	173
Surface broadcast	February	188
Surface broadcast	April	185
Side dressed	June	244
Anhydrous ammonia:		
Injected	November	196
Injected	April	196

Source: Adapted from Barber, 1974. Copyright © 1974 by National Fertilizer Solutions Association. Reprinted by permission.

[a] Relative yield was calculated based on the average yield of 48 bushels/acre when no nitrogen was applied.

Table 12-5.   Yield of winter wheat fertilized with ammonium nitrogen sources alone or with N-serve.

Treatment	Amount of nitrogen	Yield of wheat
	kg/ha	kg/ha
No applied N	0	3,696
Ammonium sulfate (F)[a]	84	4,435
Ammonium sulfate plus N-Serve (F)	84	5,107
Ammonium sulfate (S)	84	5,308
Ammonium nitrate (S)	84	5,645

Source: Adapted from Murray, 1976.

[a] F = Fall application; S = Spring application.

efficient in terms of fertilizer nitrogen recovery than was fall application. Similar results were obtained by Olson et al. (1964), where fall and spring broadcast applications are compared with summer band applications of ammonium nitrate. In this case, band applications were approximately twice as efficient as fall or spring broadcast application. Because of the combination of the vagaries of climate and the interaction of temperature and precipitation with different soil textures to provide conditions conducive to denitrification or leaching losses, it should be apparent that the application of nitrogen far in advance of the time

Table 12-6.   Effect of time and source of applied nitrogen on the yield of corn grain.

Time of N application	Source of N		
	Ammonium nitrate	Urea	Anhydrous ammonia
	--------------------yield, kg/ha[a]--------------------		
Fall plowdown (November)	6,590	6,540	6,850
Spring preplant (May)	7,650	7,650	7,830
Side-dress (June)	7,730	7,750	8,030

Source: Adapted from Stevenson and Baldwin, 1969. Copyright © 1969 by American Society of Agronomy. Reprinted by permission.

[a] Based on 15.5% moisture, the average yield for the unfertilized treatment was 5,140 kg/ha.

Table 12-7.   Response of irrigated corn to nitrogen fertilizer applied at different times[a].

Time and method of fertilizer application	Rate of N	Grain yield	Total N yield	Apparent utilization of applied N
	kg/ha	kg/ha	kg/ha	-------%-----
Control	0	4,502	86	
Fall-broadcast	45	5,286	92	25
	90	6,194	109	31
	180	6,922	130	31
Spring-broadcast	45	5,387	93	28
	90	6,294	111	34
	180	7,179	138	35
Summer-banded	45	6,216	108	58
	90	7,000	127	51
	180	7,325	146	39

Source: Adapted from Olson et al., 1964.

[a] Values are means of 14 experiments of predominantly medium to fine textured soils using ammonium nitrate as the N source.

when a plant is established enough to utilize the nitrogen will result in a low nitrogen fertilizer recovery. In some situations where climatic conditions prevent transformations of nitrogen to forms that are subject to losses, it may be possible to apply fertilizer more in advance of the time when the plant will require the nitrogen.

A feature of the plant that is related to climatic effects is the interaction of temperature on nitrogen uptake by the plant. At low temperatures, nitrogen uptake may be inhibited, as was demonstrated in the lettuce plant (Frota and Tucker, 1972). The temperatures at which this inhibition of nitrogen uptake

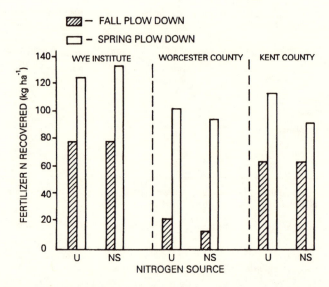

Figure 12-4 Apparent fertilizer nitrogen recovery by corn following fall or spring plow-down of urea (U) or nitrate of soda (NS) at 123 kg/ha of N for three years at locations shown, 1971. (Bandel and Rivard, 1973.)

occurs will differ among plant species. However, sufficient data is not yet available for a large enough number of plants species to allow utilization of this as a criterion for determining timing of nitrogen fertilizer applications. This is one area where additional research information would be useful. There is little or no information regarding the impact of high temperatures on nitrate uptake by plants.

Another important consideration regarding timing is the overall energy efficiency that might occur from split applications of nitrogen requiring more than one pass through the field with a tractor. Approximately 15,200 kilocalories (63 MJ) of energy are required to produce and transport 1 kg of nitrogen (6,897 kilocalories (29 MJ) per pound of nitrogen). If 14 liters of diesel fuel per hectare (approximately 1.5 gal per acre) are required in order to make an application of nitrogen, the expenditure of about 130,000 kilocalories (543 MJ) of energy per hectare in the form of diesel fuel would be required. This assumes approximately 37,000 BTU of energy per liter of diesel fuel at 0.252 kilocalories per BTU (1.05 kJ/BTU). In calculating the energy cost for the diesel fuel per hectare, the amount of fuel consumed varies with method of application, but in general surface application methods require on the order of 12 to 16 liters per hectare. If by making an additional application of nitrogen fertilizer one could reduce the total amount of nitrogen applied by 10 kg/ha, the net saving in energy would be slightly more than 22,000 kilocalories (92 MJ) per hectare. This energy saving is approximately equivalent to 0.4 kg of ammonium nitrogen or 0.6 liters of diesel fuel. This is about the break-even point for energy savings versus fertilizer efficiency. Unless one can save in excess of 10 kg/ha of nitrogen the extra energy cost of the mechanical application is of marginal benefit in terms of overall energy efficiency.

13

Energy (Organic Matter)

The soil matrix is teeming with a large population of microorganisms, estimated to be on the order of one billion organisms per cubic centimeter of soil volume. Soil organic matter and fresh organic residues incorporated into the soil provide the energy source to sustain this population of microorganisms. While the population of microorganisms may differ in a qualitative sense (there may be different species present in soils with widely different physical and chemical characteristics), in the quantitative sense they carry out the decomposition of organic materials in the soil in fundamentally the same way with, perhaps, small differences in rates from one soil to another. Many of the organisms that are essential for the transformation of nitrogen are ubiquitous and soil microbiologists have yet to find soils on earth that do not contain these microorganisms. However, there may be wide differences in the population of specific organisms, which influences the soils ability to support these transformations at a rapid rate. Where environmental factors and soil characteristics restrict the rate of transformation, changes in the soil factors and/or changes in the climate to more optimum conditions may allow microbes to function at an optimum rate. Factors that influence the aeration and energy supply will determine the rate and the type of transformations that may take place. Soil structure and texture influence water content and aeration, which along with the presence of organic matter are the predominant factors affecting the rates of nitrogen transformations.

Organic materials present in the soil as a result of decomposition of crop residues can influence the growth and development of a subsequent crop. The influence of water extracts from different plant species on the germination and growth of other plants is shown in Table 13-1. The effects of plant extracts on growth can range from inhibition to stimulation of growth and development. There is some evidence to indicate that hormones or auxins are the agents for stimulation of plant growth. In some cases, phenolic compounds that are frequently associated with soil incorporation of bark and sawdust from some tree species adversely affect plant growth and development. Other factors inhibiting plant growth and development have been associated with incorporation of readily decomposable organic materials. The rapid decomposition of organic materials can create anoxic conditions in soils, leading to the production of organic acids and sulfides that affects plant growth. These effects can be over-

Table 13-1.   Influence of water extract of different plant residues on germination
                and growth of roots and shoots of wheat seedlings.

Type of crop residue	Shoot	Root	Germination
	--------------------percent of control--------------------		
Alfalfa hay	23	0	81
Sorghum stalks	26	19	88
Sorghum roots	31	23	98
Rice straw	48	80	98
Safflower stems	53	43	98
Cornstalks	60	77	95
Wheat straw	79	93	95
Corn roots	85	108	106
Wheat roots	92	98	97

Source: Adapted from McCalla et al., 1964.

come by allowing sufficient time between incorporation and planting to permit near complete decomposition of these materials to occur.

In addition to the potential phytotoxic effects of crop residues on subsequent crops there is the impact of nitrogen content in plant material on the amount of nitrogen released to the soil environment. Several examples of release rates and nitrogen contents were discussed in Chapter 5. As pointed out, even though materials of low nitrogen content are incorporated into the soil, with time these materials will decompose to the point where nitrogen will be released.

Soils contain varying amounts of organic matter in different states of decomposition. For soils the term organic matter usually means the stable nitrogenous humic acids and similar compounds. The rapidly decomposing organic materials are most responsible for affecting soil physical changes through the development of aggregates. About 2 percent of the nitrogen contained in stable organic matter is mineralized each year. These releases of nitrogen for subsequent crop use depend upon the initial soil organic matter content and stage of decomposition of added crop residues or animal manures. The influence of two cropping patterns (winter fallow and winter crop) on nitrate distributions in the soil profile and the subsequent yield of cotton and sorghum is shown in Fig. 13-1. In Figure 13-1(A) there is a definite increase in the amount of nitrate-N present after a winter fallow, indicating that mineralization was taking place during the cooler winter months. The impact on yields and response to applied nitrogen are shown in Figs. 13-1(B) and 13-1(C). There was no significant yield increase with additional nitrogen when the soil was allowed to remain fallow over the winter. Where a winter crop had been grown there was a significant yield increase from applied nitrogen. These studies are indicative of the quantity and availability of nitrogen that may be available from mineralization of organic matter.

Sustained fertilization of a field leads to high residual soil nitrogen levels. Through the interaction of greater nitrogen uptake and production of larger quan-

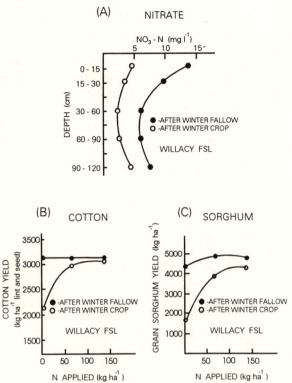

Figure 13-1   Influence of cropping system on nitrate-N concentration in a soil (A); and yield response to applied nitrogen for cotton (B), and sorghum (C) in a Willacy fine sandy loam soil (fine-loamy, mixed, hyperthermic udic Argiustolls.) (Hipp and Gerard, 1971. Copyright © 1971 by the American Society of Agronomy. Reprinted by permission.)

tities of crop residues, soil incorporation of these residues can provide increasing quantities of nitrogen to the soil by mineralization. If nitrogen becomes deficient, there is a resultant reduction in the quantities of crop residues returned to the soil and the amount of mineralized nitrogen will be reduced over time. This was the reason for the reduction in the amount of residual soil nitrate in the control plots of the experiment conducted with corn on the Davis plots by Broadbent and Carlton (1976). In the first year fertilizer nitrogen was not required in order to achieve maximum yield, but in subsequent years a higher and higher rate of applied nitrogen was required to achieve maximum yield. This is the result of the reduction of the amount of mineralizable nitrogen. It also indicates how rapidly the so-called readily decomposed organic nitrogen reserve may be depleted with time.

The presence of a crop or any organic material also influences the type and rate of nitrogen transformations that take place in the soil profile. Studies by Parkin (1987) and Staley et al. (1990) have shown that greater amounts of denitrification occurred in soils where readily decomposable carbon had been added to the soil. Rolston (1977) showed that where a crop had been grown or animal manure had been added there was greater denitrification than in fallow soil because of the increase in readily available energy supply for microorganisms to carry out denitrification. The impact of this is seen on the nitrate distribution in the soil profile under three cropping systems in Fig. 13-2. The greater the amount of available energy, the lower nitrate level found in the soil profile. This would lead to a lower efficiency of utilization of nitrogen fertilizer resources.

Irrigation management or method, insofar as it affects the water content of the soil, will have an impact on the rate of decomposition and the type of nitrogen transformations that will occur. Under saturated soil conditions, the mineralization of organic matter results in the release of ammonium to the soil environment, but because of the anoxic conditions the microorganisms that transform ammonium to nitrate are unable to function. Incorporation of organic matter into soils that are soon to become saturated would have the effect of allowing decomposition but retaining the nitrogen in ammonium form where it could not undergo denitrification losses or be subject to leaching losses. However, once the soils became aerobic then nitrification would occur at sufficiently rapid rates that large quantities of nitrogen could be changed to the nitrate form.

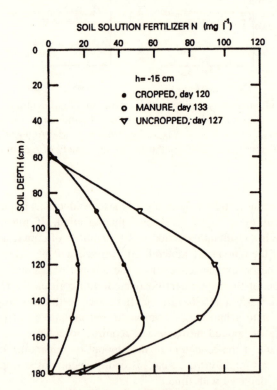

Figure 13-2 Concentration of fertilizer derived NO_3^--N (ppm N) in the soil solution as a function of depth for the cropped, manured, and uncropped plots at a soil-water pressure head of -15 cm for approximately 125 days after fertilizer application. (Rolston et al., 1977.)

Then as the soils became saturated again either through precipitation or irrigation, the potential for denitrification losses is great. As shown in Fig. 13-3, the amount of nitrogen lost by denitrification increases substantially as the soil becomes saturated. Even with flood irrigation methods, the surface few centimeters are the only portion of the soil profile that are completely saturated. Unfortunately, the surface few centimeters of soil also corresponds to that portion of the soil profile where the organic matter content is highest. This means that the energy source available for the microorganisms to carry out denitrification is generally sufficient in agricultural soils so that energy is not a limiting factor. Some manipulation of irrigation method may be desirable in order to minimize the magnitude of denitrification loss. The impact of water management on denitrification is shown in Table 5-6.

The amount of crop residue incorporated and the addition of nitrogen along with the organic material can have an influence on the response of a subsequent crop to nitrogen fertilizer applications. The effects of no fall-applied nitrogen and spring-applied nitrogen on sugar beet yields with different rates of straw incorporation are shown in Fig. 13-4. There were only slight differences between spring fertilization with or without the addition of nitrogen in the fall as discussed previously. This indicates that, if adequate time is allowed for decomposition to occur, even at a slower rate with no additional nitrogen, the organic matter will reach the point where there is net release of nitrogen into the soil environment. Consequently, one need only fertilize for the crop and it is gener-

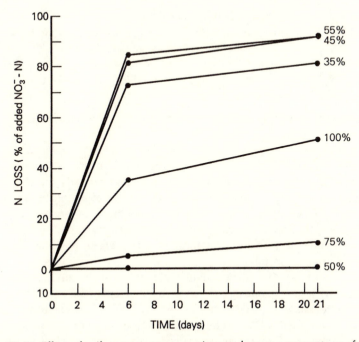

Figure 13-3 Effect of soil-water content on nitrogen loss as a percentage of water-holding capacity in soil incubated at 25°C. (Bremner and Shaw, 1958. Copyright © 1958 by Cambridge University Press. Reprinted by permission.)

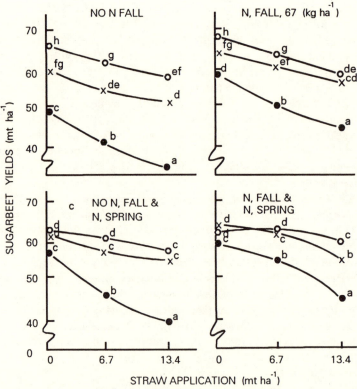

Figure 13-4 Sugarbeet yields as influenced by straw applications and nitrogen fert-ilizer. Points on the curves with different letters are different at the 95% probability level for each individual graph. MT = metric tons. (Smith et al., 1973. Copyright © 1973 by the American Society of Agronomy. Reprinted by permission.)

ally unnecessary to apply nitrogen to decompose the straw and then also fertilize for the crop.

Another feature of organic matter decomposition is the preferential utilization of the ammonium-N in the decomposition of organic matter. As shown in Fig. 13-5 (see also Fig. 5-2) ammonium-N was preferentially utilized in the decomposition of straw, which may have an adverse effect on the availability of nitrogen. Once nitrogen has become immobilized in organic form it is released back to the soil environment at a relatively low rate. It is well known from microbiological studies that most microorganisms preferentially utilize ammonium in their metabolism.

As with all other biological activities, the effect of the environment on the type of activity and the rate of reaction makes it an important consideration with respect to the fate of organic matter in soils and thus the availability of organic nitrogen. Even at temperatures close to freezing, the rate of mineralization, nitri-fication, and denitrification can be adequate to account for substantial changes in the forms of nitrogen present in the soil environment or loss from the soil environment. Because these are biological processes, at temperatures below 0°C,

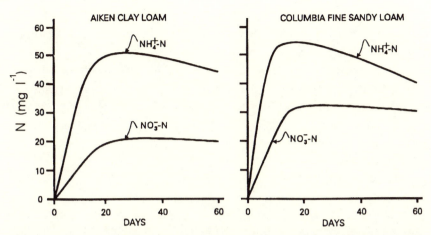

Figure 13-5 The immobilization of nitrogen from ammonium sulfate and calcium nitrate applied at a concentration of 100 ppm NO_3^--N during organic matter decomposition in an Aiken clay loam (clayey, oxic, mesic xeric Haplohults) and in a Columbia fine sandy loam (coarse-loamy, mixed, nonacid, thermic aquic Xerofluvents). (Abridged from Broadbent and Tyler, 1965. Copyright © 1965 by Kluwer. Reprinted by permission.)

these transformations essentially cease. In some irrigated areas there are sufficiently cool temperatures during periods of the year that decomposition and other biological processes cease. Consequently, this should be taken into account in managing the nitrogen supply or selecting sources of nitrogen for fertilization.

14
Summary

It should be amply apparent from the preceding discussions that agricultural production systems are exceedingly complex with numerous factors that must be considered in relation to nitrogen management in irrigated agriculture. There are several points that should be emphasized for the reader to put what has been presented into perspective. First, it should be abundantly clear that various forms of nitrogen in fertilizer do not all react in the same way in soil or water. Considerable emphasis has been placed on the differences that exist as they affect the reactions in soils, availability to plants, and the types of losses the fertilizers undergo. It is convenient to think only in terms of nitrogen when referring to all nitrogen fertilizer materials, but in order to utilize nitrogen fertilizers efficiently one must fully understand the differences between nitrogen sources and how their use is affected by system and management variables.

The crop is another feature of the agricultural production system that plays a major role in the determination of nitrogen-use efficiency and the amount of carryover. Plants are incapable of removing all of the available nitrogen from the soil–water matrix. This has a special significance in relation to environmental protection because plant demand for nitrogen and the ability of the root systems to utilize soil and fertilizer nitrogen supplies are characteristics that determine the potential for nitrogen carryover, which in turn may contribute to pollution water supplies.

The movement of water through the soil profile, the essentiality of water for growth and development, and the different mobilities of various forms of nitrogen in soils are features of the system that are stressed because of their importance to the manipulation of nitrogen supplies in order to enhance the utilization of nitrogen by plants.

When examining potential environmental degradation from nitrogen, it is important to look at both the concentration that may be reaching the water supplies and the mass emission. This is related to the movement of nitrogen with water through soils and the ability of the plant to remove only a portion of the soil nitrogen. It has been amply demonstrated that small amounts of water percolating below the root zone lead to high concentrations of nitrogen in the soil solution. Conversely, large amounts of water percolating through the root zone lead to low concentrations of nitrogen in the soil solution leaving the root

Table 14-1.   Guidelines or criteria for judging the relative sensitivity of an area to leaching of nitrate from irrigated lands.

Factor	Guidelines or criteria		
	Low sensitivity I	Medium sensitivity II	High sensitivity III
Receiving water	Not a source requiring low NO_3 concentrations.	Intermediate situations.	Multiple uses, some requiring low NO_3 concentrations.
	Already has such high NO_3 that more will do no damage.		Low dilution of drainage waters.
	High dilution of drainage waters.		No alternate supplies.
	Irrigated agriculture is an insignificant source of NO_3.		Economic impact of NO_3 leaching is high.
			Irrigated agriculture is a significant source of NO_3.
Soils	Clayey soils and soils having layers that restrict water flow, limit drainage volume and promote denitrification.	Loamy soils, intermediate in water flow characteristics.	Sandy soils having no layers that restrict water flow.
			Well aggregated soils that have high water flow features.
Crops	Require low N inputs and/or have high N use efficiencies.	Good mixture of crops requiring high N inputs with low efficiency of use with crops that are efficient and that require low N inputs.	Vegetable and fruit crops of low N use efficiency requiring high N inputs.
	Hay crops including legumes, grains, sugarbeets and grapes.		No or low acreage of highly efficient crops in the area.

Continued--.

Table 14-1 (Continued)

| Factor | Criteria or Guidelines | | |
| | Low Sensitivity | Medium Sensitivity | High Sensitivity |
	I	II	III
Irrigation	Efficient systems and management that allows low drainage volumes. Typically well-managed sprinkler or drip systems with controls.	Carefully managed surface irrigation systems where low drainage volume is expected. Mixture of efficient and inefficient systems.	Inefficient systems that promote large drainage volumes. Typically surface flow systems with long irrigation runs and large amounts of water are used.
Climate	Low rainfall that creates no leaching hazard.	Infrequent rains that occasionally promote leaching.	Heavy rains concentrated in a short period. Temperatures are sufficiently high for nitrification and winter crops are grown.

Source: Pratt, 1979c.

zone. The latter may contribute substantially greater amounts of nitrogen to water supplies.

Several of these factors have been integrated by Pratt (1979c) into guidelines for judging the likelihood of nitrates leaching to the groundwater. These guidelines for determining the sensitivity of an area to leaching losses are shown in Table 14-1.

In this book both system and management variables have been discussed from the standpoint that there is a wide range of conditions to be found for each of the variables. In combination one can expect to find a continuum of possibilities from very low to very high nitrogen-use efficiency in the agricultural production system. With a thorough knowledge of the system variables and how the management variables that are available can be used, it is possible to operate at peak efficiency for any given set of conditions. Peak efficiency will vary from one site to the next and it will change as the sytem variables change. Nevertheless, for any given set of system variables, the efficiency of nitrogen-use attainable is determined through the manipulation of management variables. The management variables were discussed in an order that permits the use of the acronym PER SITE as a cognate and to emphasize the site specificity of their application. When operating at peak efficiency, the least amount of environmental degradation that it is possible to achieve will be the result. If this is unsatisfactory then a different decision must be made; a decision that is based on socioeconomic and environmental factors, not one based on resource management factors.

Finally, in the application of the information contained in this book the reader should be guided by one overriding aspect: *principles never change, conditions change.* There are frequent debates about whether crop production and the use of nitrogen fertilizers are contributing to nitrogen pollution of our water supplies. One can find single sets of data to support either conclusion. The facts of the matter are that conditions exist under which little or no nitrogen degradation of the water supplies is occurring and there are conditions where it can be easily demonstrated that nitrogen pollution is occurring. Both situations are indeed correct and neither is the general case. It then becomes the responsibility of each person to ascertain which conditions prevail for a specific site and to apply the principles contained within this book to protect the environment and concomitantly sustain agricultural production of food and fiber.

References

Abbott, J.L. 1975. *Use Animal Manures Effectively*. Bulletin A-55. College of Agriculture, University of Arizona, Tucson, AZ.

Adams, F. and J.B. Martin. 1984. Liming effects on nitrogen use efficiency. In R.D. Hauck et al. (ed.), *Nitrogen in Crop Production*. Amer. Soc. Agron., Madison, WI, pp. 417–426.

Adriano, D.C., P.F. Pratt, S.E. Bishop, W. Brock, J. Oliver, and W. Fairbank. 1971. Nitrogen load of soil in groundwater from dairy manure. *Calif. Agric.* 25(12):12–14.

Allison, F.E. 1956. Estimating the ability of soils to supply nitrogen. *Ag. Chemicals* 11:46–48, 139.

Allison, F.E. 1961. Nitrogen transformations in soils. *Soil Crop Sci. Soc. Florida* 21:248–254.

Allison, F.E., M. Kefouver, and E.M. Roller. 1953. Ammonium fixation in soils. *Soil Sci. Soc. Amer. Proc.* 17:107–110.

Amer, F.M. 1949. Influence of oxygen percentage and moisture tension on nitrification in soils. M.S. Thesis, Iowa State College, Ames, IA.

Anonymous. 1975. Irrigation survey. *Irrigation J.* 23:15–22.

Anonymous. 1977. *Agricultural Water Use in a Dry Year*. San Joaquin County Crop Digest. A University of California Cooperative Extension Newsletter, Stockton, CA.

Anonymous. 1991. 1990 Irrigation survey. *Irrigation J.* 41:23–34.

Ardakani, M.S., H. Fluhler, and A.D. McLaren. 1976. Materials balance for a nutrient in a soil-plant system: II. Rates of nitrate uptake with sudan grass and reduction in a field. In P.F. Pratt (ed.), *Nitrates in Effluents From Irrigated Lands*. Annual Report to the National Science Foundation, Research Applied to National Needs Program. University of California, Riverside, pp. 197–226.

Ayers, R.S. and R.L. Branson (eds.). 1973. *Nitrates in the Upper Santa Ana River Basin in Relation to Groundwater Pollution*. Bulletin 861. University of California Agricultural Experiment Station, Davis, CA.

Ayers, R.S. and D.W. Westcot. 1985. *Water Quality for Agriculture*. Irrigation and Drainage Bulletin 29, Revision 1. Food and Agriculture Organization, Rome, Italy.

Azevedo, J. and P.R. Stout. 1974. *Farm Animal Manures: An Overview of their Role in the Agricultural Environment*. Manual 44. University of California Division of Agriculture Sciences, Davis, CA.

Baker, J.M. and B.B. Tucker. 1973. *Fertilizer Recommendation Guide*. Extension Facts No. 2225. Oklahoma State University, Stillwater, OK.

Bandel, V.A. and C.E. Rivard. 1973. *Influence on Corn of Fall vs Spring Nitrogen Fertilization and Plowing*. Bulletin MP 843. Maryland Agricultural Experiment Station, University of Maryland, College Station, MD.

Barber, S.A. 1974. Nitrogen efficiency Part 2: The Midwest. *Fertilizer Solutions* **5**:58–60.

Barrons, K.C. 1971. Environmental benefits of intensive crop production. *Down to Earth* **27**(3):18–22.

Baver, L.D., W.H. Gardner, and W.R. Gardner. 1972. *Soil Physics*. John Wiley, New York.

Beeson, K.C. 1941. *The Mineral Composition of Crops with Particular Reference to the Soils in Which They Were Grown*. U.S.D.A. Misc. Pub. 369. U.S. Department of Agriculture, Washington, D.C.

Berry, J.T. and N.L. Hargett. 1989. *Fertilizer Summary Data, 1988*. Bulletin Y-209. National Fertilizer Development Center, Tennessee Valley Authority, Mussel Shoals, AL.

Bertrand, A.R. and H. Kohnke. 1957. Subsoil conditions and their effects on oxygen supply and the growth of corn roots. *Soil Sci. Soc. Amer. Proc.* **21**:135–140.

Bielby, D.G., M.H. Miller, and L.R. Webber. 1973. Nitrate content of percolates from manured lysimeters. *J. Soil Water Conserv.* **28**:124–125.

Biggar, J.W. and D.R. Nielsen. 1976. The spatial variability of the leaching characteristics of a field soil. *Water Resources Research* **12**:78–84.

Biggar, J.W., D.R. Nielsen, and J.L. MacIntyre. 1976. The evaluation of soil water flux based upon field measurements. In P.F. Pratt (ed.), *Nitrates in Effluents from Irrigated Lands*. Annual Report to the National Science Foundation. Research Applied to National Needs Program. University of California, Riverside, pp. 35–60.

Black, C.A. 1957. *Soil–Plant Relationships*. John Wiley, New York.

Brazelton, R.W., K.C. Lee, S. Roy, and N.B. Akesson. 1970. *New Concepts in Aircraft Granular Applications*. A.S.A.E. Paper No. 70-657. Amer. Soc. Agric. Eng., St. Joseph, MI.

Bremner, J.M. and L.A. Douglas. 1971. Decomposition of urea phosphate in soils. *Soil Sci. Soc. Amer. Proc.* **35**:575–578.

Bremner, J.M. and K. Shaw. 1958. Denitrification in soil. *J. Agr. Sci.* **51**:22–52.

Broadbent, F.E. 1953. The soil organic matter fraction. In A.G. Norman (ed.), *Advances in Agronomy*. Academic Press, New York, pp. 153–183.

Broadbent, F.E. 1972. *Nitrogen Fixation*. Mimeographed Notes. Department of Land, Air and Water Resources, University of California, Davis, CA.

Broadbent, F.E. 1976. Field trials with isotopes, Davis site. In P.F. Pratt (ed.), *Nitrates in Effluents from Irrigated Lands*. Annual Report to the National Science Foundation, Research Applied to National Needs Program. University of California, Riverside, pp. 15–30.

Broadbent, F.E. and A.B. Carlton. 1976. Field trials with isotopes, Kearney site. In P.F. Pratt (ed.), *Nitrates in Effluents from Irrigated Lands*. Annual Report to the National Science Foundation, Research Applied to National Needs Program. University of California, Riverside, pp. 31–35.

Broadbent, F.E. and A.B. Carlton. 1979. Methodology for field trials with ^{15}N-depleted fertilizers. In P.F. Pratt (ed.), *Nitrates in Effluents from Irrigated Lands*. Final Report to the National Science Foundation, Research Applied to National Needs Program. University of California, Riverside, pp. 33–56.

Broadbent, F.E. and F.E. Clark. 1965. Denitrification. In W.V. Bartholomew and F.E. Clark (eds.), *Soil Nitrogen*. Agronomy Monograph 10. Amer. Soc. Agron., Madison, WI, pp. 344–359.

Broadbent, F.E. and T. Nakashima. 1967. Reversion of fertilizer nitrogen in soils. *Soil Sci. Soc. Amer. Proc.* **31**:648–652.

Broadbent, F.E. and T. Nakashima. 1968. Plant uptake and residual value of six tagged nitrogen fertilizers. *Soil Sci. Soc. Amer. Proc.* **32**:388–392.

Broadbent, F.E. and K.B. Tyler. 1962. Laboratory and greenhouse investigations of nitrogen immobilization. *Soil Sci. Soc. Amer. Proc.* **26**:459–462.

Broadbent, F.E. and K.B. Tyler. 1965. Effect of pH on nitrogen immobilization in two California soils. *Plant and Soil* **23**:314–322.

Broadbent, F.E., G.N. Hill, and K.B. Tyler. 1958a. Transformations and movement of urea in soils. *Soil Sci. Soc. Amer. Proc.* **22**:303–307.

Broadbent, F.E., K.B. Tyler, and G.N. Hill. 1958b. Nitrification of fertilizers. *Calif. Agric.* **12**(10):9.

Buol, S.W. 1972. Fertility capability soil classification system. In *Agronomic–Economic Research on Tropical Soils*. Annual Report for 1971 and 1972 to USAID. Dept. of Soil Science, North Carolina State University, pp. 45–50.

Buol, S.W., P.A. Sanchez, R.B. Cote, and M.A. Granger. 1975. Soil fertility capability classification. In E. Bornemisza and A. Alvarado (eds.), *Soil Management in Tropical America*. Department of Soil Science, North Carolina State University, pp. 126–140.

Busch, J.R., D.W. Fitzsimmons, G.C. Lewis, D.V. Naylor, and K.H. Yoo. 1975. *Factors Influencing the Loss of Nitrogen and Phosphorus from a Tract of Irrigated Land*. Amer. Soc. Agric. Eng. Paper 75-243.

California Department of Water Resources. 1976. Concepts of reasonable and unreasonable use. In *Proceedings of Agricultural Water Conservation Conference*, June 1976, Co-sponsored by California Department of Water Resources and the University of California, Cooperative Extension Service, pp. 245–249.

California Interagency Agricultural Information Task Force. 1977a. *Selecting an Irrigation System—Should You Change?* A California Interagency Agricultural Information Task Force Publication: Drought Tips.

California Interagency Agricultural Information Task Force. 1977b. *Common Irrigation Problems—Some Solutions*. A California Interagency Agricultural Information Task Force Publication: Drought Tips.

Carbon, F., G. Girad, and F. Ledox. 1991. Modelling of nitrogen cycle in farm land areas. *Fertilizer Research* **27**:161–199.

Carter, D.L., J.A. Bondurant, and C.W. Robbins. 1971. Water soluble NO_3-Nitrogen, PO_4-Phosphorus and total salt balance on a large irrigation tract. *Soil Sci. Soc. Amer. Proc.* **35**:331–335.

Carter, J.N., M.E. Jensen, and S.M. Bosma. 1974. Determining nitrogen fertilizer needs for sugarbeets from residual soil nitrate and mineralizable nitrogen. *Soil Sci. Soc. Amer. J.* **66**:319–323.

Carter, J.N., D.T. Westermann, M.E. Jensen, and S.M. Bosma. 1975. Predicting nitrogen fertilizer needs for sugarbeets from residual nitrate and mineralizable nitrogen. *J. Amer. Soc. Sugarbeet Technologists* **18**:232–244.

Chandra, P. and W.B. Bollen. 1961. Effects of nabam and mylone on nitrification, soil respiration, and microbial numbers in four Oregon soils. *Soil Sci.* **29**:387–393.

Christensen, S., S. Simkins, and J.M. Tiedje. 1990. Spatial variation in denitrification: Dependency of activity centers on the soil environment. *Soil Sci. Soc. Amer. J.* **54**:1608–1613.

Clay, D.E., G.L. Malzer, and J.L. Anderson. 1990. Ammonia volatilization from urea as influenced by soil temperature, soil water content and nitrification and hydrolysis inhibitors. *Soil Sci. Soc. Amer. J.* **54**:263–266.

Conn, E.E. and P.F. Stumpf. 1963. *Outlines of Biochemistry*. John Wiley, New York.

Conrad, J.P. 1942. Enzymatic and microbial concepts of urea hydrolysis in soils. *J. Amer. Soc. Agron.* **34**:1102–1113.

Cooper, G.S. and R.L. Smith. 1963. Sequence of products formed during denitrification in some diverse Western soils. *Soil Sci. Soc. Amer. Proc.* **27**:659–662.

Cope, J.T. 1975. Vetch for green manure—time to reconsider? *Highlights of Agric. Research, Auburn University, Alabama* **22**:3.

Corvinka, V., W.J. Chancellor, R.J. Coffelt, R.G. Curley, and J.B. Dobie. 1974. *Energy Requirements for Agriculture in California*. Report of a Joint Study by the California Department of Food and Agriculture and the University of California, Davis.

Cox, W.J. and H.M. Reisenauer. 1973. Growth and ion uptake by wheat supplied nitrogen as nitrate or ammonium or both. *Plant and Soil* **38**:363–380.

Day, J.C. and G.L. Horner. 1987. *U.S. Irrigation, Extent and Economic Importance*. Agricultural Information Bulletin 523. U.S. Department of Agriculture, Economic Resarch Service, Washington, D.C.

Davidson, J.M., P.S.C. Rao, and R.E. Jessup. 1978. Behavior of nitrogen in the plant root zone: A critique of computer simulation modeling for nitrogen in irrigated crop lands. In D.R. Nielsen and J.G. MacDonald (eds.), *Nitrogen and the Environment*. Academic Press, New York, pp. 131–143.

Delwiche, C.C. 1970. The nitrogen cycle. *Scientific American* **233**:137–146.

Delwiche, C.C. and G.E. Likens. 1977. Biological response to fossil fuel combustion products. In W. Stumm (ed.), *Global Chemical Cycles and Their Alterations by Man*. Dahlem Konferengen, Berlin, pp. 73–88.

Donohue, S.J., C.L. Rhykerd, D.A. Holt, and C.H. Noller. 1973. Influence of N fertilization and N carryover on yield and N concentration of *Dactylis Glomerota* L. *Agron. J.* **65**(4):671–674.

Dow, A.I., D.W. James, and C.E. Nelson. 1969. *Interpretation of Soil Test Nitrogen for Irrigated Crops in Central Washington*. Coop. Ext. Serv. EM 3076. Washington State Univ., Pullman, WA.

Dunnigan, E.D., R.A. Phelan, and C.L. Mondart, Jr. 1976. Surface runoff losses of fertilizer elements. *J. Environ. Qual.* **5**:339–342.

Dutt, G.R., M.J. Shaffer, and W.J. Moore. 1972. *Computer Simulation Model of Dynamic Bio-Physicochemical Processes in Soils*. Tech. Bulletin 196. Ariz. Agric. Exp. Stn., University of Ariz., Tucson.

Engler, R.M., D.A. Antie, and W.H. Patrick, Jr. 1976. Effect of dissolved oxygen on redox potential and nitrate removal in flooded swamp and marsh soils. *J. Environ. Qual.* **5**:230–235.

Erdman, L.W. 1959. *Legume Inoculation: What It Is, What It Does*. U.S.D.A. Farmers Bulletin, 2003. United States Department of Agriculture, Washington, D.C.

Erickson, R.O. 1976. Modeling of plant growth. *Annual Rev. Plant Physiol.* **27**:407–434.

Ernst, J.W. and H.F. Massey. 1960. The effects of several factors on volatilization of ammonia formed from urea in the soil. *Soil Sci. Soc. Amer. Proc.* **24**:87–90.

FAO. 1989. *FAO Annual Production Handbook, 1988*. FAO Statistic Series No. 88, Vol. 42. Food and Agriculture Organization, Rome, Italy.

Fenn, L.B. and R. Escarzaga. 1976. Ammonia volatilization from surface applications of ammonium compounds on calcareous soils: V. Soil water content and method of nitrogen application. *Soil Sci. Soc. Amer. J.* **40**:537–541.

Fenn, L.B. and R. Escarzaga. 1977. Ammonia volatilization from surface applications of ammonium compounds on calcareous soils: VI. Effects of initial soil-water content and quantity of applied water. *Soil Sci. Soc. Amer. J.* **41**:358–363.

Fenn, L.B. and D.E. Kissel. 1973. Ammonia volatilization from surface applications of ammonium compounds on calcareous soils: I. General theory. *Soil Sci. Soc. Amer. J.* **37**:855–859.

Fenn, L.B. and D.E. Kissel. 1974. Ammonia volatilization from surface applications of ammonium compounds on calcareous soils: II. Effects of temperature and rate of ammonium nitrogen application. *Soil Sci. Soc. Amer. Proc.* **38**:606–610.

Fenn, L.B. and D.E. Kissel. 1976. The influence of cation exchange capacity and depth of incorporation on ammonia volatilization from ammonium compounds applied to calcareous soils. *Soil Sci. Soc. Amer. J.* **40**:394–398.

Ferguson, A.H. 1959. Movement of soil water inferred from moisture content measurements by gamma-ray absorption. Ph.D. Dissertation, Washington State University, Pullman, WA.

Ferguson, A.H. and W.H. Gardner. 1962. Water content measurement in soil columns by gamma ray absorption. *Soil Sci. Soc. Amer. Proc.* **26**:11–14.

Fitzsimmons, D.W., G.C. Lewis, K.H. Lindeborg, J.R. Busch, D.V. Naylor, and D.H. Forteir. 1975. *Effects of On-Farm Water Management Practices on Water Quality in the Boise Valley.* Idaho Agric. Exp. Stn. Report on Corp. of Engineers Contract No. DACW 68-74-C-0071. University of Idaho, Moscow.

Fitzsimmons, D.W., C.E. Brockway, J.R. Busch, G.L. Lewis, G.M. McMaster, and C.W. Berg. 1977. On farm methods for controlling sediment and nutrient losses. In J.P. Law and G.V. Skogerboe (eds.), *Proceedings of National Conference on Irrigation Return Flow Quality Management.* Colorado State University, Fort Collins, CO, pp. 183–192.

Frederick, L.R. 1956. The formation of nitrate from ammonium nitrogen in soils: I. Effect of temperature. *Soil Sci. Soc. Amer. Proc.* **20**:496–500.

Free, G.R., G.M. Browning, and G.S. Musgrave. 1940. *Relative Infiltration and Related Physical Characteristics of Certain Soils.* U.S.D.A. Technical Bulletin 729. U.S. Department of Agriculture, Washington, D.C.

Fried, M., K.K. Tanji, and R.M. Van De pol. 1976. Simplified long term concepts for evaluating leaching of nitrogen from agricultural land. *J. Environ. Qual.* **5**(2):197–200.

Frissel, M.J. and J.A. van Veen. 1978. Computer simulation of nitrogen behavior in soil, a critique. In D.R. Nielsen and J.G. MacDonald (eds.), *Nitrogen and the Environment*, Vol. I. Academic Press, New York, pp. 145–162.

Frota, J.N.E. and T.C. Tucker. 1972. Temperature influence on ammonium and nitrate absorption by lettuce. *Soil Sci. Soc. Amer. Proc.* **36**:97–100.

Gardner, B.R. and W.D. Pew. 1972. *Response of Fall Grown Head Lettuce to Nitrogen Fertilizer.* Tech. Bulletin 199. University of Arizona Agric. Exp. Stn., Tucson, AZ.

Gardner, B.R. and W.D. Pew. 1974. *Response of Spring Grown Head Lettuce to Nitrogen Fertilizer.* Tech. Bulletin 210. University of Arizona Agric. Exp. Stn., Tucson, AZ.

Gardner, B.R. and T.C. Tucker. 1967. Nitrogen effects on cotton: II. Soil and petiole analysis. *Soil Sci. Soc. Amer. Proc.* **31**(6):785–791.

Gardner, B.R., M.D. Openshaw, and T.C. Tucker. 1976. *Wheat Fertilizer Recommendations.* University of Arizona Agricultural Engineering and Soil Science Newsletter. University of Arizona, Tucson, AZ.

Gardner, E.H. 1971. *OSU Soil Test for Nitrate Nitrogen.* Circular 770. Oregon Cooperative Extension Service, Oregon State University, Corvallis, OR.

Gass, W.B., G.A. Peterson, R.D. Hauck, and R.A. Olson. 1971. Recovery of residual nitrogen by corn (*Zea mays* L.) from various soil depths as measured by ^{15}N tracer techniques. *Soil Sci. Soc. Amer. Proc.* **35**:290–294.

Gasser, J.K.R. 1965. *Some Processes Affecting Nitrogen in the Soil*. Tech. Bulletin No. 15. Rothamsted Experiment Station, UK.

Geraghty, J.J., D.W. Miller, F. Van Der Leeden, and F.L. Troise (eds.), 1973. *Water Atlas*. Water Information Center, Inc., Port Washington, NY.

Ghosh, B.P. and R.J. Burris. 1950. Utilization of nitrogenous compounds by plants. *Soil Sci.* **70**:187–203.

Gilford, R.O., G.L. Ashcroft, and M.D. Magnuson. 1967. *Probability of Selected Precipitation Amounts in the Western Region of the United States: Section for Colorado*. Tech. Bulletin 7-8. Western Regional Research publication. Nevada Agric. Exp. Sta., University of Nevada, Reno, NV.

Goring, C.A.I. 1962. Control of nitrification of ammonium fertilizers and urea by 2-chloro-6-(trichloromethyl)pyridine. *Soil Sci.* **93**:431–439.

Gossett, F.L. and N.K. Whittlesey. 1976. *Cost of Reducing Sediment and Nitrogen Outflows from Irrigated Farms in Central Washington*. Bulletin 824. Washington State University College of Agriculture Research Center, Pullman, WA.

Graham, P.H. 1976. Identification and classification of root nodule bacteria. In P.S. Nutman (ed.), *Symbiotic Nitrogen Fixation in Plants*. Cambridge University Press, Cambridge, pp. 99–112.

Groot, J.J.R. and P. de Willigen. 1991. Simulation of nitrogen balance in soil and a winter wheat crop. *Fertilizer Research* **27**:261–272.

Hagin, J. and A. Amberger. 1974. *Contribution of Fertilizers and Manures to the N- and P-Load of Waters. A Computer Simulation*. Report to Deutsche Forschungs Gemeinschaft from Technion-Israel Institute of Technology, Soil and Fertilizer Div., Haifa, Israel.

Hahn, G.J. and S.S. Shopiro. 1967. *Statistical Models in Engineering*. John Wiley, New York.

Hansen, S., H.E. Jensen, N.E. Nielsen, and H. Svendsen. 1990. *DAISY: A Soil–Plant System Model. Danish Simulation Model for Transformation and Transport of Energy and Matter in the Soil–Plant–Atmosphere System*. The National Agency for Environmental Protection, Copenhagen, Denmark.

Hardesty, J.O. 1967. Watch the salt content. *Farm Chemicals* **29**(10):4.

Hardy, R.W.F. and A.H. Gibson. 1977. *A Treatise on Dinitrogen Fixation: Section IV. Agronomy and Ecology*. John Wiley, New York.

Harridine, F. and H. Jenny. 1958. Influence of parent material and climate on texture and nitrogen and carbon contents of virgin California soil. I. Texture and nitrogen content of soils. *Soil Sci.* **85**:235–243.

Hattori, A. 1957. Studies on the metabolism of urea and other nitrogenous compounds in *Chlorella ellipsoidea*. I. Assimilation of urea and other nitrogenous compounds by nitrogen-starved cells. *J. Biochem.* **44**:253–273.

Heermann, D.F. and H.H. Shull. 1970. *Effective Precipitation of Various Application Depths*. ASAE Paper 70-735. Amer. Soc. Ag. Engineers, St Joseph, MI.

Henderson, D.W., W.C. Bianchi, and L.D. Doneen. 1955. Ammonia loss from sprinkler jets. *Agric. Eng.* **36**:398–399.

Hills, F.J. 1976. Estimating the amount of fertilizer nitrogen required for a sugarbeet crop. In *Advances in Agronomy and Range Science*. University of California Division of Agriculture, Special Publication 3204, pp. 3–4.

Hipp, B.W. and C.J. Gerard. 1971. Influence of previous crop and nitrogen mineralization on crop response to applied nitrogen. *Agron. J.* **63**:583–586.

Israelson, O.W. and V.E. Hansen. 1967. *Irrigation Principles and Practices*, 3rd edn. John Wiley, New York.

James, D.W. 1971. *Soil Fertility Relationships of Sugarbeets in Central Washington: Nitrogen*. Tech. Bulletin 68. Washington Agricultural Experiment Station, Washington State Univ., Pullman, WA.

James, S.W., C.E. Nelson, and A.R. Halvorson. 1967. *Soil Testing for Residual Nitrates as a Guide for Nitrogen Fertilization of Sugarbeets*. Circular 480. Washington State University College of Agriculture.

Jensen, D. and J. Pesek. 1962. Inefficiency of fertilizer use resulting from nonuniform spatial distribution: I. Theory. *Soil Sci. Soc. Amer. Proc.* **26**:170–174.

Jensen, M.E., J.N. Carter, and B.J. Ruffing. 1965. *Irrigation Water Management on Sugarbeets*. Annual Report, U.S.D.A. A.R.S. Snake River Conservation Research Center, Kimberly, ID.

Johnson, H., L. Bergstrom, P.-E. Jasson and K. Paustian. 1987. Simulated nitrogen dynamics and losses in a layered agricultural soil. *Agric. Ecosystems and the Environment* **18**:333–356.

Jones, P. and C.G. Painter. 1974. *Tissue Analysis: A Guide to Nitrogen Fertilization of Idaho Russet Burbank potatoes*. Current Information Series No. 240. University of Idaho, College of Agriculture, Moscow, ID.

Jordan, J.H., Jr., W.H. Patrick, Jr., and W.H. Willis. 1967. Nitrate reduction by bacteria isolated from waterlogged Crowley soil. *Soil Sci.* **104**:129–133.

Jung, J. 1972. Factors determining the leaching of nitrogen from soils, including some aspects of maintenance of water quality. *Qual. Plant Mater. Veg.* **XXI**(4):343–366.

Kerbs, L.D., J.P. Jones, W.L. Thiessen, and F.P. Parks. 1973. Correlation of soil test nitrogen with potato yields. *Commun. Soil Sci. Plant Analysis* **4**:269–278.

Kersebaum, K.C. and J. Richter. 1991. Modelling nitrogen dynamics in a plant-soil system with a simple model for advisory purposes. *Fertilizer Research* **27**:273–281.

Kilmer, V.J., J.W. Gillian, J.F. Lutz, R.T. Joyce, and C.D. Eklund. 1974. Nutrient losses from fertilized grass watersheds in Western North Carolina. *J. Environ. Qual.* **3**:214–219.

Kirby, E.A. and K. Mengel. 1967. Ionic balance in different tissues of the tomato plant in relation to nitrate, urea or ammonium nitrogen. *Plant. Physiol.* **42**:6–14.

Klemmedson, J.O. and H. Jenny. 1966. Nitrogen availability in California soils in relation to precipitation and parent material. *Soil Sci.* **102**:215–222.

Kliewer, M.W. and J.A. Cook. 1974. Arginine levels in grape canes and fruits as indicators of nitrogen status of vineyards. *Amer. J. Enol. Viticult.* **25**:111–115.

Knutson, J.D., Jr., R.G. Curley, E.B. Roberts, R.H. Hagen, and V. Cervinka. 1977. Energy for irrigation. *Calif. Agric.* **31**(6):46–57.

Krantz, B.A. and M.D. Miller. 1968. Plant nutrition and crop quality. *Agrichemical West* **11**(12):15–18.

Kresge, C.B. and D.P. Satchell. 1959. Gaseous loss of ammonia from nitrogen fertilizers applied to soils. *Agron. J.* **52**:104–107.

Kroontje, W. and W.R. Kehr. 1956. Legume top and root yields in the year of seeding and subsequent barley yields. *Agron. J.* **48**:127–131.

Krull, D.L. and D.L. Clark. 1977. An assessment of irrigation efficiencies and drainage return flows from the Welton-Mohawk Division of the Gila Project. In J.P. Law and G.V. Skogerboe (eds.), *Proceedings of National Conference on Irrigation Return Flow Quality Management*. Colorado State University, Fort Collins, pp. 335–348.

Lafolie, F. 1991. Modelling water flow, nitrogen transport and root uptake including physical non-equilibrium and optimization of the root water potential. *Fertilizer Research* **27**:215–231.

Lance, J.C. 1972. Nitrogen removal by soil mechanisms. *J. Water Pollut. Control Fed.* **44**:1352–1361.

Lang, A.L., J.W. Pendleton, and G.H. Dungan. 1956. Influence of populations and nitrogen levels on yield and protein and oil contents of nine corn hybrids. *Agron. J.* **48**:284–289.

Lehane, J.J. and W.J. Staple. 1953. Water retention and availability in soils related to drought resistance. *Canadian J. of Agricultural Sci.* **33**:265–273.

Loomis, R.S. 1976. Agricultural systems. *Scientific American* **235**:98–105.

Lorenz, O.A., B.L. Weir, and J.C. Bishop. 1972. Effect of controlled-released nitrogen fertilizers on yield and nitrogen adsorption by potatoes, cantaloupe and tomatoes. *J. Amer. Soc. Hort. Sci.* **97**(3):334–337.

Ludwick, A.E., P.N. Soltanpour, and J.O. Reuss. 1977. Nitrate distribution and variability in irrigated fields. *Agron. J.* **69**:710–713.

Lund, L.J. and R.A. Elliott. 1976. Nitrate and chloride leaching as related to soil profile characteristics. In P.F. Pratt (ed.), *Nitrates in Effluents from Irrigated Lands.* Annual Report to the National Science Foundation, Research Applied to National Needs Program. University of California, Riverside, pp. 127–140.

Luthin, J.N. 1970. Movement of water through soils. In *Relationship of Agriculture to Soil and Water Pollution.* Proceedings of a Cornell University Conference on Agricultural Waste Management, 1970, Rochester, New York, pp. 21–28.

Madison, R.J. and J.O. Brunett. 1984. Overview of the occurrence of nitrate in groundwater of the United States. In *National Water Summary*, Water Supply Paper 2275. U.S. Geological Survey, Reston, VA, pp. 93–105.

Martin, J.P. and H.D. Chapman. 1951. Volatilization of ammonium from surface fertilized soils. *Soil Sci.* **71**:25–34.

Mayberry, K.S. 1977. *Fertilizer Burn.* Imperial County Agricultural Briefs. University of California Agricultural Extension Publication, November 1977, pp. 4–5.

Mayberry, K. and R.S. Rauschkolb. 1975. Nitrogen uptake in midwinter lettuce. *Calif. Agric.* **29**(3):6–7.

Maynard, D.N. and A.V. Barker. 1969. Studies on the tolerance of plants to ammonium nutrition. *J. Amer. Hort. Sci.* **94**:235–239.

Mayurak, A.P. and E.C. Conrad. 1966. Changes in content of total nitrogen and organic matter in three Nebraska soils after seven years of cropping treatments. *Agron. J.* **58**:85–88.

McAuliffe, C., D.S. Chamblee, H. Uribe-Arango, and W.W. Woodhouse. Jr. 1958. Inorganic nitrogen and fixation by legumes using N. *Agron. J.* **50**:334–337.

McCalla, T.M., W.D. Guenzi, and F.A. Norstadt. 1964. Phytotoxic substances in stubble mulching. In *8th International Congress of Soil Science*, Bucharest, Romania, pp. 933–943.

McElroy, A.D., S.Y. Chin, J.W. Nebgen, A. Aleti, and F.W. Bennett. 1976. *Loading Functions for Assessment of Water Pollution From Non-point Sources.* EPA-600/2-76-151. U.S. Environmental Protection Agency.

McLaren, A.D. 1969. Steady state studies of nitrification in soils: Theoretical considerations. *Soil Sci. Soc. Amer. Proc.* **33**:273–275.

McLaren, A.D. 1976. Comments on nitrate reduction in unsaturated soil. *Soil Sci. Soc. Amer. Proc.* **40**:698–699.

McVickar, M.H. 1967. *Using Commercial Fertilizers.* The Interstate Printers and Publishers, Inc., Danville, IL.

McVickar, M.H., G.L. Bridger, and L.B. Nelson. 1963. *Fertilizer Technology and Usage*. Soil Sci. Soc. Amer., Madison, WI.

Meek, B.O., L.B. Grass, and A.J. MacKenzie. 1969. Applied nitrogen losses in relation to oxygen status of soils. *Soil Sci. Soc. Amer. Proc.* **33**:575–578.

Meek, B., L. Chesnin, W. Fuller, R. Miller, and D. Turner. 1975. *Guidelines for Manure Use and Disposal in the Western Region, USA*. Bulletin 814. Washington State University, College of Agriculture, Pullman, WA.

Mehran, K. and K.K. Tanji. 1974. Computer modeling of nitrogen transformations in soils. *J. Environ. Qual.* **3**:391–395.

Menzies, J.D. and R.L. Chaney. 1974. Waste characteristics. In *Factors Involved in Land Application of Agricultural and Municipal Wastes*. U.S.D.A. Agricultural Research Service Report, Beltsville, MD, pp. 18–50.

Middleton, J.E., E. Roberts, D.W. James, T.A. Cline, B.L. McNeal, and B.L. Carlile. 1975. *Irrigation and Fertilizer Management for Efficient Crop Production on a Sandy Soil*. Bulletin 811. Washington State University, College of Agriculture, Pullman, WA.

Mikkelsen, D.S. and D.C. Finfrock. 1957. Availability of ammoniacal nitrogen to lowland rice as influenced by fertilizer placement. *Agron. J.* **49**:296–300.

Mikkelsen, D.S. and W.H. Patrick, Jr. 1968. Fertilizer use in rice. In L.C. Nelson (ed.), *Changing Patterns in Fertilizer Use*. Soil Sci. Soc. Amer., Madison, WI, pp. 403–432.

Miller, N.H.J. 1906. The amount and composition of drainge through unmanured and uncropped land, Barnfield. *Rothamsted J. Agric. Sci.* **1**:377–399.

Miller, R.J. and D.W. Wolfe. 1977. Nitrogen inputs and outputs: A valley basin study. In D.R. Nielsen and J.G. MacDonald (eds.), *Nitrogen and the Environment*. Academic Press, New York, pp. 291–300.

Miller, R.J., D.E. Rolston, R.S. Rauschkolb, and D.W. Wolfe. 1976. Drip application of nitrogen is efficient. *Calif. Agric.* **30**(11):16–18.

Mitchell, R.L. 1970. *Crop Growth and Culture*. Iowa State University Press, Ames, IA.

Miyamoto, S., J. Ryan, and J.L. Stroehlein. 1975. Sulfuric acid for the treatment of ammoniated irrigation water: I. Reducing ammonia volatilization. *Soil Sci. Soc. Amer. Proc.* **39**:544–548.

Moore, F.D., III and P.N. Soltanpour. 1974. Slow release fertilizers and nitrification suppressants as nitrogen management tools for lettuce. *Agronomy Abstracts* **66**:152.

Munns, D.N. 1977. Mineral nutrition and the legume symbiosis. In R.W.F. Hardy and A.H. Gibson (eds.), *Treatise on Dinitrogen Fixation, Section IV*. John Wiley, New York, pp. 353–391.

Murray, G.A. 1976. *N-Serve and its Potential Use in Northern Idaho*. Current Information Series No. 313. University of Idaho College of Agriculture.

National Commission on Water Quality. 1976. *Report to the Congress by the National Commission on Water Quality*. Washington, D.C.

National Plant Food Institute. 1966. *Hunger Signs in Crops*. National Plant Food Institute, Washington, D.C.

Nelson, C.E., R.E. Early, and A. Mortensen. 1965. *Nitrogen Fertilization of Corn in Relation to Soil Nitrate Nitrogen*. Circular 453. Washington Agricultural Experiment Station, Washington State University, Pullman, WA.

Nelson, D.W. 1973. Losses of fertilizer nutrients in surface runoff. *Fertilizer Solutions* **17**(May–June):10–13.

Nielsen, D.R., R.D. Jackson, J.W. Cory, and D.D. Evans (eds.). 1970. *Soil Water*. Western Regional Research Technical Committee on Water Movement in Soils, W68.

American Society of Agronomy and the Soil Science Society of America, Madison, WI.

Nielsen, D.R., J.W. Biggar, and K.T. Erh. 1974. Spatial variability of field-measured soil-water properties. *Hilgardia* **42**:215–260.

Nielsen, R.F. and L.A. Banks. 1960. A new look at nitrate movement. Utah State University Experiment Station. *Farm and Home Sci.* **21**:2–3.

Nutman, P.S. 1965. Symbiotic nitrogen fixation. In W.V. Bartholomew and F.E. Clark (eds.), *Soil Nitrogen.* Amer. Soc. of Agron., Madison, WI, pp. 360–383.

Olson, R.A. 1976. *Recent Nitrogen Research in Nebraska.* Abstracts of Joint U.S.D.A.-A.R.S. and the Fertilizer Institute Nitrate Research Review Conference, Fort Collins, CO.

Olson, R.A., K.D. Frank, and A.F. Dreier. 1964. Controlling losses of fertilizer nitrogen from soils. In *Proc. 8th International Congress of Soil Science*, Bucharest, Romania, pp. 1023–1032.

Onken, A.B., C.W. Wendt, R.S. Hargrove, and O.C. Wilke. 1977. Relative movement of bromide and nitrate in soils under three irrigation systems. *Soil Sci. Soc. Amer. J.* **41**:50–52.

Openshaw, M.D. 1972. *Arizona Sugarbeet Fertilizer Recommendations.* Agri-File Q-34. University of Arizona Cooperative Extension, Tucson, AZ.

Openshaw, M.S., B.R. Gardner, W.D. Pew, and T.C. Tucker. 1973. *Lettuce Fertilizer Recommendations.* Agri-File Q-263. University of Arizona Cooperative Extension, Tucson, AZ.

Osterli, P.P. and J.L. Meyer. 1976. *Nitrogen Fertilization and Pollution Potential.* A University of California Cooperative Extension County Report, Stanislaus County, Modesto, CA.

Overrein, L.N. 1968. Lysimeter studies on tracer nitrogen in forest soil. 1. Nitrogen losses by leaching and volatilization after addition of urea-N^{15}. *Soil Sci.* **106**:280–290.

Overrein, L.N. 1969. Lysimeter studies on tracer nitrogen in forest soil. 2. Comparative losses of nitrogen through leaching and volatilization after additions of urea-, ammonium- and nitrate-N^{15}. *Soil Sci.* **107**:149–159.

Overrein, L.N. 1971. Isotopic studies of nitrogen in forest soil. I. Relative losses of nitrogen through leaching during a period of forty months. Norwegian Forest Research Institute, Vollebekk, Norway. *Meddeleser Norske Skogforsoksveen* **114**:264–280.

Papendick, R.I. and J.F. Parr. 1966. Retention of anhydrous ammonia by soil: III. Dispensing apparatus and resulting ammonia distribution. *Soil Sci.* **102**:193–201.

Papendick, R.I., J.F. Parr, S. Smith, and R.W. Smiley. 1971. Nitrification inhibition in soil: II. Evaluation of anhydrous-potassium azide solutions in Eastern Washington. *Soil Sci. Soc. Amer. Proc.* **35**:579–582.

Parker, D.T. and W.E. Larson. 1962. Nitrification as affected by temperature and moisture content of mulched soils. *Soil Sci. Soc. Amer. Proc.* **26**:238–242.

Parkin, T.A. 1987. Soil moisture as a source of denitrification variability. *Soil Sci. Soc. Amer. J.* **51**:1194–1199.

Patrick W.H., Jr. 1960. Nitrate reduction rates in submerged soil as affected by redox potential. In *Proc. 7th International Congress of Soil Science*, Vol. II, Madison, WI, pp. 494–500.

Patrick, W.H., Jr. and S. Gotoh. 1974. The role of oxygen in nitrogen loss from flooded soils. *Soil Sci.* **118**:78–81.

Patrick, W.H., Jr., J.J. Peterson, and F.T. Turner. 1968. Nitrification inhibitors for lowland rice. *Soil Sci.* **105**:103–105.

Pilot, L. and W.H. Patrick, Jr. 1972. Nitrate reduction in soils: Effect of soil moisture tension. *Soil Sci.* **114**:312–316.

Porter, L.K. and A.R. Grable. 1969. Fixation of atmospheric nitrogen by nonlegumes in wet mountain meadows. *Agron. J.* **61**:521–523.

Powers, F.F. 1968. Mineralization of nitrogen in grass roots. *Soil Sci. Soc. Amer. Proc.* **32**:673–674.

Pratt, P.F. (ed.). 1979a. *Nitrate in Effluents from Irrigated Lands.* Final Report to the National Science Foundation, Research Applied to National Needs Program. University of California, Riverside.

Pratt, P.F. 1979b. Estimated leaching and denitrification losses of nitrogen in a four-year field trial with manures. In P.F. Pratt (ed.), *Nitrate in Effluents from Irrigated Lands.* Final Report to the National Science Foundation, Research Applied to National Needs Program. University of California, Riverside, pp. 321–354.

Pratt, P.F. 1979c. Integration, discussion and conclusions. In P.F. Pratt (ed.), *Nitrate in Effluents from Irrigated Lands.* Final Report to the National Science Foundation, Research Applied to National Needs Program. University of California, Riverside, pp. 719–758.

Pratt, P.F., F.E. Broadbent, and J.P. Martin. 1973. Using organic wastes as nitrogen fertilizers. *Calif. Agric.* **27**(6):10–13.

Pratt, P.F., L.J. Lund, and J.E. Warneke. 1976. Nitrogen losses in relation to soil profile characteristics. In P.F. Pratt (ed.), *Nitrates in Effluents from Irrigated Lands.* Annual Report to the National Science Foundation, Research Applied to National Needs Program. University of California, Riverside, pp. 141–166.

Rao, D.N. and D.S. Mikkelsen. 1976. Effect of rice straw incorporation on rice plant growth and nutrition. *Agron. J.* **68**:752–755.

Rao, P.S.C., J.M. Davidson, and R.E. Jessup. 1981. Simulation of nitrogen behavior in the root zone of cropped land areas receiving organic wastes. In M.J. Frissel and H. van Veen (eds.), *Simulation of Nitrogen Behaviour of Soil Water Systems.* Centre for Agric. Publications and Documentation, Wageningen, The Netherlands, pp. 81–95.

Rauschkolb, R.S. 1968. Nitrogen absorption and utilization by *Gossypium hirsutum* as influenced by nitrogen source. Ph.D. Dissertation, University of Arizona, Dept. of Agricultural Chemistry and Soils, Tucson, AZ.

Rauschkolb, R.S. 1974a. An appraisal of the energy requirement for fertilizers in California. University of California Cooperative Extension Service, *Soil and Water Newsletter* No. 21:3–5.

Rauschkolb, R.S. 1974b. Biological fixation of nitrogen. In *Symposium on Nitrogen Especially as Related to Agricultural Production and Water Quality.* California Dept. of Water Resources, Sacramento, CA and the University of California, Davis, Kearney Foundation.

Rauschkolb, R.S., A.L. Brown, J. Quick, J.D. Prato, R.E. Pelton, and F.R. Kegel. 1974a. Rapid tissue testing for evaluating nitrogen nutritional status of corn. *Calif. Agric.* **28**(6):10–12.

Rauschkolb, R.S., A.L. Brown, R.L. Sailsbery, J. Quick, J.D. Prato, and R.E. Pelton. 1974b. Rapid tissue testing for evaluating nitrogen nutritional status of sorghum. *Calif. Agric.* **28**(6):12–13.

Rauschkolb, R.S., R.D. Bottell, J. Vanderhill, W.N. Helphenstine, and R. Chavariia. 1975. Land application of fruit and vegetable cannery by-products. University of California Cooperative Extension Service, *Soil and Water Newsletter* No. 26:3–4.

Ray, H.E., T.C. Tucker, and L.R. Amburgery. 1964. *Soil and Petriole Analysis Can*

Pinpoint Cotton's Nitrogen Needs. Folder 97. University of Arizona College of Agriculture, Tucson, AZ.

Reed, A.D., W.E. Wildman, W.S. Seyman, R.S. Ayers, J.D. Prato, and R.S. Rauschkolb. 1973. Soil recycling of cannery wastes. *Calif. Agric.* **27**(3):6–9.

Reed, A.D., J.L. Meyers, F.K. Aljibury, and A.W. Marsh. 1977. *Irrigation Costs.* Leaflet No. 2875. University of California, Division of Agricultural Services.

Reddy, K.R. and W.H. Patrick, Jr. 1976. Effect of frequent changes in aerobic and anaerobic conditions on redox potential and nitrogen loss in a flooded soil. *Soil Biol. Biochem.* **8**:491–495.

Reuss, J.O., P.N. Soltanpour, and A.E. Ludwick. 1977. Sampling distribution of nitrates in irrigated fields. *Agron. J.* **69**:588–592.

Rible, J.M., P.A. Nash, P.F. Pratt, and L.J. Lund. 1976. Sampling the unsaturated zone of irrigated land for reliable estimates of nitrates concentrations. *Soil Sci. Soc. Amer. J.* **40**(4):566–570.

Rijtema, R.J., C.W.J. Roest, and J.G. Kroes. 1990. *Formulation of the Nitrogen and Phosphate Behavior in Agricultural Soils, the ANIMO Model.* Report 30. The Winard Staring Centre, Wageningen, The Netherlands.

Roberts, S., A.W. Richards, and W.H. Weaver. 1976. *Evaluating Measurements of Soil Nitrate, Mineralizable Nitrogen and Nitrate in Sugarbeet Petioles as Guides to Fertilization.* Annual Report of Washington State University, Irrigated Agriculture Research and Extension Center, Prosser, WA.

Rolston, D.E. 1977. Measuring nitrogen loss from denitrification. *Calif. Agric.* **31**(1):12–13.

Rolston, D.E. 1978. Volatile losses of nitrogen from soils. In *Nitrogen Management in Irrigated Agriculture.* A National Conference sponsored by U.S.-E.P.A., N.S.F. and the University of California, Sacramento, CA, pp. 169–193.

Rolston, D.E., D.A. Goldhamer, D.L. Hoffman, and D.W. Toy. 1977. Field measured flux of volatile denitrification products as influenced by soil-water content and organic carbon source. In J.P. Law, Jr. and G.V. Skogerboe (eds.), *Proceedings of National Conference on Irrigation Return Flow Quality Management,* Colorado State University, Ft. Collins, CO, pp. 55–61.

Roy, R.N. and B.C. Wright. 1973. Sorghum growth and nutrient uptake in relation to soil fertility: I. Dry matter accumulation patterns, yield, and N content of grain. *Agron. J.* **65**:709–711.

Roy, R.N. and B.C. Wright. 1974. Sorghum growth and nutrient uptake in relation to soil fertility: II. N, P, and K uptake patterns by various plant parts. *Agron. J.* **66**:5–10.

Rubins, E.J. and F.E. Bear. 1942. Carbon–nitrogen ratios in organic fertilizer materials in relation to the availability of their nitrogen. In *Annual Report of New Jersey Agricultural Experiment Station,* Rutgers, NJ, pp. 411–423.

Sabey, B.R. 1968. The influence of nitrification suppressants on the rate of ammonium oxidation in Midwestern USA field soils. *Soil Sci. Soc. Amer. Proc.* **32**:675–679.

Sabey, B.R., L.R. Frederick, and W.V. Bartholomew. 1969. The formation of nitrate from ammonium nitrogen in soils: IV. Use of the delay and maximum rates phases for making quantitative predictions. *Soil Sci. Soc. Amer. Proc.* **33**:276–278.

Sailsbery, R.L. 1981. *Nitrogen Top Dressing of Anza Wheat—A Four Year Summary of the Effects on Yield, Yellowbelly, and Protein in Glenn County.* Farm Adviser Office, Glenn Co., CA.

Sailsbery, R.L. and F.J. Hills. 1976. *Sugarbeet Fertilization.* A University of California Cooperative Extension County Publication, Glenn Co., CA.

Sain, P. and F.E. Broadbent. 1977. Decomposition of rice straw in soils as affected by some management factors. *J. Environ. Qual.* **6**:96–100.

Sanchez, P.A., W. Couto, and S.W. Buol. 1982. The fertility capability soil classification system: Interpretation, applicability and modifications. *Geoderma* **27**:283–309.

Sartain, J.B. 1976. *Proper Fertilizer Management Saves Energy.* Energy Conservation Fact Sheet EC-27. Florida Cooperative Extension Service, University of Florida, Gainesville, FL.

Schade, R.O., L.K. Stromberg, E.A. Yeary, B.B. Burlinzome, R.G. Curley, B.A. Krantz, W.E. Martin, and V.P. Osterli. 1962. *Fertilizer Application Costs to Grower.* Publication AXT-62. University of California, Cooperative Extension Service.

Scherty, D.L. and D.A. Miller. 1972. Nitrate-N accumulation in the soil profile under alfalfa. *Agron. J.* **64**:660–664.

Selim, H.M. and I.K. Iskandar. 1981. A model for predicting nitrogen behavior in slow and rapid infiltration systems. In I.K. Iskandar (ed.), *Modeling Wastewater Renovation by Land Treatment.* John Wiley, New York, pp. 479–507.

Selim, H.M., P. Kanchamasut, R.S. Mansell, L.W. Zelanzy, and J.M. Davidson. 1974. Phosphorus and chloride movement in a spodosol. *Soil and Crop Sci. Soc. Florida* **34**:19–23.

Shaffer, M.J., R.W. Ribbens, and C.W. Huntly. 1977. *Detailed Return Flow Salinity and Nutrient Simulation Model: V. Prediction of Mineral Quality of Irrigation Return Flow.* EPA-600/2-77-179e. U.S. Environmental Protection Agency.

Shankaracharaya, N.B. and B.V. Mexta. 1969. Evaluation of loss of nitrogen by ammonia volatilization from soil fertilized with urea. *J. Indian Soc. Soil Sci.* **17**:423–430.

Sharratt, W.J. 1976. Nitrogen fertilizer manufacture during times of feedstock shortage. University of California Cooperative Extension Service, *Soil and Water Newsletter* No. 28:2–9.

Shearer, M.N. 1969. Uniformity of water distribution from sprinklers as it is related to the application of agricultural chemicals, water storage efficiency, sprinkler system capacity and power requirements, A communication problem. Paper at Pacific Northwest Amer. Soc. of Agric. Eng. Meeting, Vancouver, B.C., October.

Simpson, J.R. and J.R. Freney. 1967. The fate of labelled mineral nitrogen after addition to three pasture soils of different organic contents. *Australian J. Agric. Res.* **18**:613–623.

Smika, D.E., D.F. Heerman, H.R. Duke, and A.R. Batchelder. 1977. Nitrate-N percolation through irrigated sandy soil as affected by water management. *Agron. J.* **69**:623–626.

Smith, J.H., C.L. Douglas, and M.J. LeBaron. 1973. Influence of straw application rates, plowing dates, and nitrogen applications on yield and chemical composition of sugarbeets. *Agron. J.* **65**(5):797–800.

Smith, J.H., C.W. Robbins, J.A. Bondurant, and C.W. Hoyden. 1976. Treatment of potato processing waste water on agricultural land: Water and organic loading, and the fate of applied nutrients. *Proceedings 8th Annual Waste Management Conference, Land as a Waste Management Alternative,* Rochester, New York, pp. 769–781.

Sokai, H. 1959. Effect of temperature on nitrification in soils. *Soil and Plant Food* **4**:159–162.

Solley, W.B., C.F. Merk, and R.R. Pierce. 1988. *Estimated Water Use in the United States.* Circular 1004. U.S. Geological Survey, Alexandria, VA.

Soltanpour, P.N. 1969. Effect of nitrogen, phosphorous and zinc placement on yield and composition of potatoes. *Agron. J.* **61**:288–289.

Staley, T.E., W.H. Caskey, and D.G. Boyer. 1990. Soil denitrification and nitrification potentials during the growing season relative to tillage. *Soil Sci. Soc. Amer. J.* **54**:1602–1608.

Stanford, G. and S.J. Smith. 1972. Nitrogen mineralization potentials of soils. *Soil Sci. Soc. Amer. Proc.* **36**:465–472.

Stanford, G., J.N. Carter, and S.J. Smith. 1974. Estimates of potentially mineralizable soil nitrogen, based on short-term incubations. *Soil Sci. Soc. Amer. Proc.* **38**:99–102.

Stevenson, C.K. and C.S. Baldwin. 1969. Effect of time and method of nitrogen application and source of nitrogen on the yield and nitrogen content of corn (*Zea Mays* L.). *Agron J.* **61**(3):381–384.

Stewart, B.A., D.D. Johnson, and L.K. Porter. 1963a. The availability of fertilizer nitrogen immobilized during decomposition of straw. *Soil Sci. Soc. Amer. Proc.* **27**:656–659.

Stewart, B.A., L.K. Porter, and D.D. Johnson. 1963b. Immobilization and mineralization of nitrogen in several organic fractions of the soil. *Soil Sci. Soc. Amer. Proc.* **27**:302–306.

Stewart, J.I. 1971. Soil water use by non-irrigated corn. University of California Cooperative Extension Service, *Soil and Water Newsletter* No. 11:1–3.

Stewart, J.I. 1975. *Irrigation in California*. Report to the State Water Resources Control Board. Department of Land, Air, and Water Resources, University of California, Davis.

Stewart, W.D.P. 1966. *Nitrogen Fixation in Plants*. The Athlone Press, University of London.

Steyn, P.L. and C.C. Delwiche. 1970. Nitrogen fixation by nonsymbiotic microorganisms in some California soils. *Environ. Sci. Technol.* **4**:1122–1128.

Stout, P.R. and R.G. Burau. 1967. The extent and significance of fertilizer buildup in soils as revealed by vertical distributions of nitrogen matter between soils and underlying water reservoirs. In *Agriculture and the Quality of Our Environment*. American Association for the Advancement of Science, Washington, D.C., pp. 283–310.

Tanji, K.K. and S.K. Gupta. 1978. Computer simulation modeling for nitrogen in irrigated croplands. In D.R. Nielsen and J.G. MacDonald (eds.), *Nitrogen and the Environment: Vol. I. Nitrogen Behavior in Field Soil*. Academic Press, New York. pp. 79–130.

Tanji, K.K., J.W. Biggar, G.L. Horner, R.J. Miller, and W.O. Pruitt. 1976. *Irrigation Tailwater Management*. Annual Report on U.S.E.P.A. Grant #R803603-01. Department of Land, Air, and Water Resources. University of California, Davis and U.S.D.A. Economic Research Service, Davis.

Tanji, K.K., M. Mehran, and S.K. Gupta. 1981. Water and nitrogen fluxes in the root zone of irrigated maize. In M.J. Frissel and H. van Veen (eds.), *Simulation of Nitrogen Behavior of Soil–Plant Systems*. Centre for Agricultural Publications and Documentation, Wageningen, The Netherlands, pp. 51–66.

Terman, G.L. and C.M. Hunt. 1964. Volatilization losses of nitrogen from surface applied fertilizers as measured by crop response. *Soil Sci. Soc. Amer. Proc.* **28**:667–672.

Thorne, D.W. and H.B. Peterson. 1954. *Irrigated Soils: Their Fertility and Management*. Blakiston Co. Inc., New York.

Tillotson, W.R., C.W. Robbins, R.J. Wagenet, and R.J. Hanks. 1980. *Soil Water, Solute and Plant Growth Simulation*. Bulletin 502. Utah Agric. Exp. Stn., Utah State Univ., Logan, UT.

Tisdale, S.L. and W.L. Nelson. 1966. *Soil Fertility and Fertilizers* (2nd edn). Macmillan, New York.

Tisdale, S.L., W.L. Nelson, and J.D. Beaton. 1985. *Soil Fertility and Fertilizers* (4th edn). Macmillan, New York.

Treshow, M. 1970. *Environment and Plant Response*. McGraw-Hill, New York.

Turner, G.O., L.E. Warren, and F.G. Androssen. 1962. Effect of 2-chloro-6-(trichloromethyl)pyridine on the nitrification of ammonium fertilizers in field soils. *Soil Sci.* **94**:270–273.

Tyler, K.B., F.E. Broadbent, and V. Kondo. 1958. Nitrogen movement in simulated cross sections of field soil. *Agron. J.* **50**:626–628.

Tyler, K.B., F.E. Broadbent, and G.N. Hill. 1959. Nitrification of fertilizers. *Calif. Agric.* **13**(7):10.

Ulrich, A. and F.J. Hills. 1973. Plant analysis as an aid in fertilizing crops: Part I. Sugar beets. In R.C. Dinauer (ed.), *Soil Testing and Plant Analysis*. Soil Sci. Soc. Amer., Madison, WI, pp. 271–288.

U.S.D.A. 1983. *Commercial Fertilizer Consumption for Year Ending June 30, 1983*. U.S.D.A., Washington, D.C.

U.S.D.A. Economic Research Service. 1992. *Economic Indicators of the Farm Sector: Production and Efficiency Statistics, 1990*. Resources and Technology Division, Economic Research Service, U.S.D.A., ECIFS 10-3, Washington, D.C.

U.S. Department of Commerce. 1990. *1987 Census of Agriculture, Volume 2, Part 1, Agricultural Atlas of the United States*. AC87-S-1. U.S. Department of Commerce, Washington, D.C.

Vanderlip, R.L. and H.E. Reeves. 1972. Growth stages of sorghum [*Sorghum bicolor* (L.) Moench.] *Agron. J.* **64**:13–16.

Verrecken, H., M. Vanclooster, and M. Swerts. 1990. A model for the estimation of nitrogen leaching and regional applicability. In Merckx, R., H. Vereecken, and K. Vlassak (eds.), *Fertilizer and the Environment*. Leuven University Press, Leuven, Belgium, pp. 250–263.

Viets, F.B., Jr. and S.R. Aldrich. 1973. Crop production: Sources of nitrogenous compounds and methods of control. In *Nitrogenous Compounds in the Environment*. EPA-SAB-73-001. Hazardous Materials Advisory Committee Report to the U.S.E.A.P.A., pp. 67–93.

Viets, F.G. and R.H. Hageman. 1971. *Factors Affecting the Accumulation of Nitrate in Soil Water and Plants*. U.S.D.A./A.R.S., Agriculture Handbook No. 413. U.S.D.A., Washington, D.C.

Vines, H.N. and R.T. Wedding. 1960. Some effects of ammonia on plant metabolism and a possible mechanism for ammonia toxicity. *Plant Physiol.* **35**:820–823.

Virtanen, A.I. and H. Linkola. 1946. Organic nitrogen compounds in nitrogen nutrition for higher plants. *Nature* **158**:515–517.

Volk, G.M. 1961. Gaseous loss of ammonia from surface applied nitrogenous fertilizers. *J. Agric. Food Chem.* **9**:280–283.

Volk, G.M. 1965. Transformations and mobility of band-placed urea. *Soil Crop Sci. Soc. Florida* **25**:202–211.

Wagenet, R.J. and J.L. Hutson. 1989. *LEACHM: Leaching Estimation and Chemistry Model: a Process Model of Water and Solute Movement, Transformations, Plant Uptake and Chemical Reactions in the Unsaturated Zone*. Continuum Vol. 2, Version 2. Water Resources Institute, Cornell Univ., Ithaca, NY.

Warrick, A.W., G.J. Mullen, and D.R. Nielsen. 1977. Predictions of soil water flux based upon field-measured soil-water properties. *Soil Sci. Soc. Amer. J.* **41**:14–18.

Watt. B.K. and A.L. Merrill. 1963. *Composition of Food: Raw, Processed and Prepared*. U.S.D.A. Handbook #8. U.S.D.A., Washington, D.C.

Wendt, C.W., A.B. Onken, O.C. Wilke, and R.D. Lacewell. 1977. *Effects of Irrigation*

Methods on Groundwater Pollution by Nitrates and Other Solutes. EPA-600/2-76-291. Report to U.S. Environmental Protection Agency, Robert S. Kerr Research Laboratory, Ada, OK.

Went, F.W. 1957. Climate and agriculture. *Scientific American* **196**:82–94.

Western Fertilizers Handbook. 1980. 6th Edition. Produced by the Soil Improvement Committee of the California Fertilizer Association, Interstate Printers and Publishers, Inc., Danville, IL.

Westerman, R.L. and T.C. Tucker. 1974. Effect of salts and salts plus nitrogen-15-labelled ammonium chloride on mineralization of soil nitrogen, nitrification and immobilization. *Soil Sci. Soc. Amer. Proc.* **38**:602–605.

Whistler, F.D., B. Acock, D.N. Baker, R.E. Fye, H.F. Hodges, J.R. Lambert, H.E. Lemmon, J.M. McKinion, and V.R. Reddy. 1986. Crop simulation models in agronomic systems. *Adv. Agron.* **40**:141–238.

White, E.M., C.R. Krueger, and R.A. Moore. 1976. Changes in total N organic matter, available P, and bulk densities of a cultivated soil 8 years after tame pastures were established. *Agron. J.* **68**:581–583.

Wildman, W.E. 1969. Water penetration: Where is the restricting layer? University of California Cooperative Extension Service, *Soil and Water Newsletter* No. 7(Fall):1–4.

Wildman, W.E. and K.D. Gowans. 1975. *Soil Physical Environment and How it Affects Plant Growth.* Leaflet 2280. University of California, Cooperative Extension Service.

Winely, C.L. and C.L. San Clemente. 1969. The effect of pesticides on nitrate oxidation by *Nitrobacter agilis. Bacteriol. Proc.* **4**:175–179.

Wittwer, S.H., M.J. Bukorac, and H.B. Tukey. 1963. Advances in foliar feeding of plant nutrients. In M.H. McVickar, G.L. Bridger, and L.B. Nelson (eds.), *Fertilizer Technology and Usage.* Soil Sci. Soc. Amer., Madison, WI, pp. 429–448.

Index